Electrical installation calculations

VOLUME 2

A. J. Watkins
B.Sc., B.Sc.(Aston), C.Eng., M.I.E.E.

FIFTH EDITION
prepared by

Russell K. Parton
Formerly Head of the Department of Electrical and Motor Vehicle Engineering,
The Reid Kerr College, Paisley

A member of the Hodder Headline Group
LONDON

First published in Great Britain in 1957
Fifth edition in 1999 by
Arnold, a member of the Hodder Headline Group,
338 Euston Road, London NW1 3BH

http://www.arnoldpublishers.com

© 1998 R K Parton

British Library Cataloguing in Publication Data
A catalogue record for this book is available from the British Library

ISBN: 0 340 73185 0
  2 3 4 5 6 7 8 9 10

Commissioning Editor: Sian Jones
Production Editor: Wendy Rooke
Production Controller: Priya Gohil
Cover Design: Terry Griffith

Typeset in 10/14pt New Baskerville ITC
by Academic and Technical, Bristol.
Printed and bound in Great Britain by
Redwood Books, Trowbridge, Wiltshire

What do you think about this book? Or any other Arnold title?
Please send your comments to feedback.arnold@hodder.co.uk

ML

Electrical installation
calculations
Volume 2

Contents

Preface to the fifth edition

This book is the second of a series of three volumes intended for students of electrical installation work. It is essentially a book of examples to co-ordinate the electrical installation technology, electrical science and the associated calculations of courses following the syllabuses of the City and Guilds and also installation courses under the auspices of BTec and Scotvec. Volumes 1 and 2 should satisfy the needs of City and Guilds Course 236 Parts 1 and 2 and serve as an introduction to Volume 3, which is devoted to the advanced level of course material for City and Guilds Course 236 Course C. Representative examples are worked and a selection of problems, with answers, is provided to help the student to practise the techniques involved.

A small number of multiple choice questions are included, and to answer these the student should work through the question in the usual way and then indicate which of the answers corresponds to the result produced. It should be noted that each of the incorrect answers given is the result of using a wrong technique or a miscalculation. A student selecting the wrong answer should therefore check that he or she is using the correct method or correct calculation.

In this the fifth edition, the content has been revised to take account of the adoption by the British Standards Institution of the Institution of Electrical Engineers' Wiring Regulations, Sixteenth Edition to form BS 7671:1992.

This book takes into account Amendments No. 1, 1994 and No. 2, 1997 to British Standard 7671. Readers are advised to have as additional reference material one of the site-type electrician's guides, e.g. the *IEE On-site Guide*. It will also be of advantage to have reference to a 'full' copy of BS 7671 when working through certain of the calculations.

Certain material from the other two volumes has been included where it was felt to be essential to enable readers to study the subject further.

The author and publishers are grateful to the British Standards Institution and to the Institution of Electrical Engineers for permission to make use of data from their publications. Gratitude is also expressed to the City and Guilds for permission to use various questions from past examination papers, in certain cases slight re-wording was necessary to satisfy modern terminology and requirements. The two Institutions and the City and Guilds, however, accept no responsibility for the interpretation by the author of their specific requirements or indeed for the answers to the various questions.

R.K.P.

Kilmacolm

1998

Alternating-current circuit calculations

In a.c. circuits (Fig. 1), the current is limited by the *impedance* (Z). Impedance is measured in ohms, and

voltage = current (amperes) × impedance (ohms)

$$U = I \times Z$$

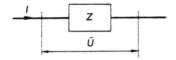

Fig. I

EXAMPLE I The current through an impedance of $24\,\Omega$ is $6\,A$. Calculate the voltage drop.

$$U = I \times Z$$
$$= 6 \times 24$$
$$= 144\,\text{V}$$

EXAMPLE 2 The current in a fluorescent-lamp circuit is $1.5\,A$ when the voltage is $230\,V$. Determine the impedance of the circuit.

$$U = I \times Z$$
$$\therefore \qquad 230 = 1.5 \times Z$$
$$\therefore \qquad Z = \frac{230}{1.5}$$
$$= 153.3\,\Omega$$

EXAMPLE 3   An a.c. contactor coil has an impedance of $300\,\Omega$. Calculate the current it will take from a $400\,V$ supply.

$$U = I \times Z$$

$$\therefore \quad 400 = I \times 300$$

$$\therefore \quad I = \frac{400}{300}$$

$$= 1.333$$

$$= 1.33\,A \text{ (correct to three significant figures)}$$

EXERCISE 1

1. Complete the following table:

Volts (a.c.)			230	400	100	25		230	
Current (A)	0.1	15	0.5		0.01		180	25
Impedance (Ω)	100	15		1000		0.05		25

2. Complete the following table:

Current (A)	1.92	3.84	18.2		7.35	4.08		8.97
Volts (a.c.)		7.5		230	107		400	235
Impedance (Ω)	2.45		12.4	96.3		56	96	

3. Complete the following table:

Impedance (Ω)	232	850		0.125		1050	129	
Volts (a.c.)		230	400	26.5	0.194		238	245
Current (A)	0.76		0.575		0.0065	0.436		0.056

4. A mercury vapour lamp takes 2.34 A when the mains voltage is 237 V. Calculate the impedance of the lamp circuit.

5. An inductor has an impedance of 365 Ω. How much current will flow when it is connected to a 400 V a.c. supply?

6. A coil of wire passes a current of 55 A when connected to a 120 V d.c. supply but only 24.5 A when connected to a 110 V a.c. supply. Calculate (a) the resistance of the coil, (b) its impedance.

7. Tests to measure the impedance of an earth fault loop were made in accordance with BS 7671 and the results for five different installations are given below. For each case, calculate the value of the loop impedance.

	Test voltage, a.c. (V)	Current (A)
(a)	9.25	19.6
(b)	12.6	3.29
(c)	7.65	23.8
(d)	14.2	1.09
(e)	8.72	21.1

8. The choke in a certain fluorescent-luminaire fitting causes a voltage drop of 150 V when the current through it is 1.78 A. Calculate the impedance of the choke.

9. Complete the following table:

Volts (a.c.)	61.1		153	193	
Current (A)	2.3	4.2		7.35	9.2
Impedance (Ω)		25	25		25

Plot a graph showing the relationship between current and voltage. From the graph, state the value of the current when the voltage is 240 V.

10. The alternating voltage applied to a circuit is 230 V and the current flowing is 0.125 A. The impedance of the circuit is

(a) 5.4 Ω (b) 1840 Ω (c) 3.5 Ω (d) 184 Ω

11. An alternating current of 2.4 A flowing in a circuit of impedance 0.18 Ω produces a voltage drop of

(a) 0.075 V   (b) 13.3 V   (c) 0.432 V   (d) 4.32 V

12. When an alternating e.m.f. of 150 V is applied to a circuit of impedance 265 Ω, the current is

(a) 39 750 A   (b) 1.77 A   (c) 5.66 A   (d) 0.566 A

INDUCTIVE REACTANCE

In alternating-current circuits which set up significant magnetic fields, there is opposition to the current in addition to that caused by the resistance of the wires. This additional opposition is called *inductive reactance*. In this section we shall consider only circuits whose resistance is so low that we may ignore it; the only opposition to the current is then the inductive reactance.

Inductive reactance $X_L = 2\pi f L$ ohms

where L is the inductance of the coil or circuit in henrys (H) and f is the supply frequency in hertz (Hz). Also (Fig. 2),

$$U_L = I \times X_L$$

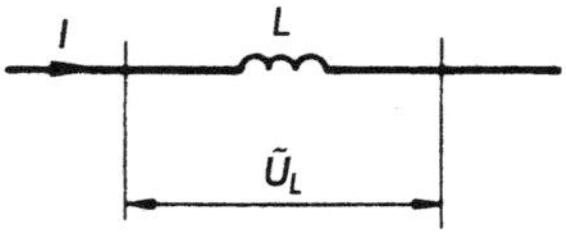

Fig. 2

EXAMPLE 1   Calculate the inductive reactance of a coil of 0.01 H inductance when connected to a 50 Hz supply.

$$\begin{aligned} X_L &= 2\pi f L \\ &= 2 \times 3.14 \times 50 \times 0.01 \\ &= 314 \times 0.01 \\ &= 3.14\,\Omega \end{aligned}$$

EXAMPLE 2   An inductor is required which will cause a voltage drop of 200 V when the current through it is 2 A at 50 Hz. Calculate the value of the inductor.

$$U_L = I \times X_L$$
$$\therefore \quad 200 = 2 \times X_L$$
$$\therefore \quad X_L = \frac{200}{2}$$
$$= 100\,\Omega$$
$$X_L = 2\pi f L$$
$$\therefore \quad 100 = 314 \times L$$
$$\therefore \quad L = \frac{100}{314}$$
$$= 0.319\,\text{H}$$

EXERCISE 2

1. Calculate the inductive reactance of a coil having an inductance of 0.015 H when a 50 Hz current flows in it.

4

2. A coil is required to have an inductive reactance of $150\,\Omega$ on a $50\,\text{Hz}$ supply. Determine its inductance.

3. Complete the following table:

Inductance (H)	0.04			0.12	0.008	
Frequency (Hz)	50	50				60
Reactance (Ω)		50	36	4.5	57	

4. A coil of negligible resistance causes a voltage drop of $98\,\text{V}$ when the current through it is $2.4\,\text{A}$ at $50\,\text{Hz}$. Calculate (a) the reactance of the coil, (b) its inductance.

5. A reactor allows a current of $15\,\text{A}$ to flow from a $230\,\text{V}$ $50\,\text{Hz}$ supply. Determine the current which will flow at the same voltage if the frequency changes to (a) $45\,\text{Hz}$, (b) $55\,\text{Hz}$. Ignore the resistance.

6. Calculate the inductive reactance of coils having the following values of inductance when the supply frequency is $50\,\text{Hz}$.

   (a) $0.012\,\text{H}$ **(d)** $350\,\mu\text{H}$

   (b) $0.007\,\text{H}$ **(e)** $0.045\,\text{H}$

   (c) $0.45\,\text{mH}$

7. Determine the inductances of the coils which will have the following reactances to a $50\,\text{Hz}$ supply:

   (a) $300\,\Omega$ **(d)** $125\,\Omega$

   (b) $25\,\Omega$ **(e)** $5\,\Omega$

   (c) $14.5\,\Omega$

8. The inductance of a coil can be varied from $0.15\,\text{H}$ to $0.06\,\text{H}$. Plot a graph to show how the inductive reactance varies with changes in inductance. Assume a constant frequency of $50\,\text{Hz}$.

9. A reactor has a constant inductance of $0.5\,\text{H}$ and it is connected to a supply of constant voltage $100\,\text{V}$ but whose frequency varies from 25 to $50\,\text{Hz}$. Plot a graph to show how the current through the coil changes according to the frequency. Ignore the resistance of the coil.

10. Calculate the voltage drop across a $0.24\,\text{H}$ inductor of negligible resistance when it carries $5.5\,\text{A}$ at $48\,\text{Hz}$.

11. An inductor of $0.125\,\text{H}$ is connected to an a.c. supply at $50\,\text{Hz}$. Its inductive reactance is

   (a) $39.3\,\Omega$ **(b)** $0.79\,\Omega$ **(c)** $0.025\,\Omega$ **(d)** $393\,\Omega$

12. The value in henrys of an inductor which has an inductive reactance of $500\,\Omega$ when connected in an a.c. circuit at frequency $450\,\text{Hz}$ is

   (a) $1.77\,\text{H}$ **(b)** $14 \times 10^{6}\,\text{H}$ **(c)** $0.177\,\text{H}$ **(d)** $0.071 \times 10^{-6}\,\text{H}$

CAPACITIVE REACTANCE

When a capacitor is connected to an a.c. supply, the current is limited by the *reactance* of the capacitor (X_C).

$$\text{Capacitive reactance } X_C = \frac{10^6}{2\pi fC} \text{ ohms}$$

where C is the capacitance of the capacitor measured in microfarads (μF) and f is the supply frequency in hertz (Hz). Also (Fig. 3),

$$U_C = I \times X_C$$

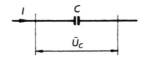

Fig. 3

EXAMPLE 1   Calculate the reactance of a $15\,\mu$F capacitor to a 50 Hz supply.

$$X_C = \frac{10^6}{2\pi fC}$$

$$= \frac{10^6}{2 \times 3.14 \times 50 \times 15}$$

$$= \frac{10^6}{314 \times 15} = 212\,\Omega$$

EXAMPLE 2   A power-factor-improvement capacitor is required to take a current of 10 A from a 230 V 50 Hz supply. Determine the value of this capacitor.

$$U_C = I \times X_C$$

$$\therefore \quad 230 = 10 \times X_C$$

$$\therefore \quad X_C = \frac{230}{10}$$

$$= 23\,\Omega$$

$$X_C = \frac{10^6}{2\pi fC}$$

where $2\pi f = 314$ when $f = 50$ and $\pi = 3.14$

$$\therefore \quad 23 = \frac{10^6}{314 \times C}$$

$$\frac{1}{23} = \frac{314 \times C}{10^6}$$

$$\therefore \quad C = \frac{10^6}{314 \times 23}$$

$$= 138.5\,\mu\text{F}$$

EXERCISE 3

1. Determine the reactance of each of the following capacitors to a 50 Hz supply. (Values are all in microfarads.)

   (a) 60   (d) 150   (g) 250   (j) 75

   (b) 25   (e) 8   (h) 95

   (c) 40   (f) 12   (i) 16

2. Calculate the value of capacitors which have the following reactances at 50 Hz. (Values are all in ohms.)

   (a) 240   (d) 4.5   (g) 45   (j) 72

   (b) 75   (e) 36   (h) 400

   (c) 12   (f) 16   (i) 30

3. Calculate the value of a capacitor which will take a current of 25 A from a 230 V 50 Hz supply.

4. A capacitor in a certain circuit is passing a current of 0.2 A and the voltage drop across it is 100 V. Determine its value in microfarads. The frequency is 50 Hz.

5. A 20 μF capacitor is connected to an alternator whose output voltage remains constant at 150 V but whose frequency can be varied from 25 to 60 Hz. Draw a graph to show the variation in current through the capacitor as the frequency changes over this range.

6. Calculate the voltage drop across a 5 μF capacitor when a current of 0.25 A at 50 Hz flows through it.

7. In order to improve the power factor of a certain installation, a capacitor which will take 15 A from the 230 V supply is required. The frequency is 50 Hz. Calculate the value of the capacitor.

8. In one type of twin-tube fluorescent fitting, a capacitor is connected in series with one of the tubes. If the value of the capacitor is $7\,\mu F$, the current through it is 0.8 A, and the supply is at 50 Hz, determine the voltage across the capacitor.

9. A machine designed to work on a frequency of 60 Hz has a power-factor-improvement capacitor which takes 12 A from a 110 V supply. Calculate the current the capacitor will take from the 110 V 50 Hz supply.

10. A capacitor takes a current of 16 A from a 400 V supply at 50 Hz. What current will it take if the voltage falls to 380 V at the same frequency?

11. A $22\,\mu F$ capacitor is connected in an a.c. circuit at 50 Hz. Its reactance is
    (a) $0.000\,145\,\Omega$ (b) $6912\,\Omega$ (c) $6\,912\,000\,\Omega$ (d) $145\,\Omega$

12. The value in microfarads of a capacitor which has a capacitive reactance of $100\,\Omega$ when connected to a circuit at 50 Hz is
    (a) $31.8\,\mu F$ (b) $318\,\mu F$ (c) $0.000\,031\,8\,\mu F$ (d) $0.0314\,\mu F$

IMPEDANCE IN SERIES CIRCUITS

Impedance is the name given to the combined effect of resistance (R) and reactance (X_L or X_C). It is measured in ohms.

For series circuits, impedance (Z) is given by

$$Z^2 = R^2 + X^2$$

where $X = X_L$ for a resistive-inductive circuit

$X = X_C$ for a resistive-capacitive circuit

and $X = X_C - X_L$ or $X_L - X_C$ (the larger minus the smaller) for a circuit with resistance, capacitance, and inductance

An inductor coil will always possess both inductance and resistance and, although the inductance and resistance cannot be physically separated, it is convenient for the purpose of calculating impedance, current, etc. to show them separately in a circuit diagram (Fig. 4).

What has been done is to assume that the inductor consists of a coil of inductance L but negligible resistance in series with a resistor R whose inductance is negligible. The impedance Z is

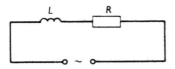

Fig. 4

then given by

$$Z^2 = R^2 + X_L{}^2$$

or $\qquad Z = \sqrt{(R^2 + X_L{}^2)}$

(It is left as a mental exercise for the reader to explain why we choose to show R and L in series and not in parallel.)

EXAMPLE 1   A coil has a resistance of $8\,\Omega$ and an inductance of $0.08\,\text{H}$. Calculate its impedance to a $50\,\text{Hz}$ supply.

$$\text{Inductive reactance } X_L = 2\pi f L$$
$$= 2\pi f \times 0.08$$
$$= 314 \times 0.08$$
$$= 25.12\,\Omega$$
$$Z^2 = R^2 + X_L{}^2$$
$$= 8^2 + (25.12)^2$$
$$= 64 + 630$$
$$= 694$$
$$\therefore \qquad Z = \sqrt{694} = 26.34$$
$$= 26.3\,\Omega$$

EXAMPLE 2   A coil passes a current of $20\,\text{A}$ when connected to a $230\,\text{V}$ d.c. supply but only $10\,\text{A}$ when connected to a $230\,\text{V}$ a.c. supply. Calculate the inductance of the coil. Take the supply frequency as $50\,\text{Hz}$.

On d.c., $\qquad U = I \times R$
$$230 = 20 \times R$$
$$\therefore \qquad R = \frac{230}{20}$$
$$= 11.5\,\Omega \text{ (resistance)}$$

On a.c., $\quad U = I \times Z$

$$230 = 10 \times Z$$

$\therefore \qquad Z = \dfrac{230}{10}$

$$= 23\,\Omega \text{ (impedance)}$$

and $\qquad Z^2 = R^2 \times X_L{}^2$

$\therefore \qquad 23^2 = 11.5^2 + X_L{}^2$

$\therefore \qquad X_L{}^2 = 23^2 - 11.5^2$

$$= 529 - 132.25$$

$$= 396.75$$

$\therefore \qquad X_L = \sqrt{396.75}$

$$= 19.92\,\Omega$$

$\qquad X_L = 2\pi f L$

$\therefore \qquad 19.92 = 2 \times 3.14 \times 50 \times L$

$\therefore \qquad L = \dfrac{19.92}{314}$

$$= 0.06344$$

$$= 0.063\,\text{H}$$

EXAMPLE 3   A $100\,\Omega$ resistor is wired in series with a capacitor of unknown value to the 230 V 50 Hz mains, and a current of 1.6 A flows (Fig. 5). Calculate the value of the capacitor in microfarads.

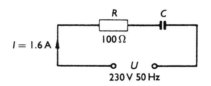

Fig. 5

First find the impedance Z:

$$U = I \times Z$$

$$230 = 1.6 \times Z$$

$$\therefore \quad Z = \frac{230}{1.6}$$

$$= 143.75\,\Omega$$

$$Z^2 = R^2 + X_C{}^2$$

$$\therefore \quad 143.75 = 100^2 \times X_C{}^2$$

$$\therefore \quad X_C{}^2 = 143.75^2 - 100^2$$

$$= (143.75 - 100)(143.75 + 100)$$

(Note alternative method, 'difference of two squares')

$$= 43.75 \times 243.75$$

$$= 10\,664$$

$$\therefore \quad X_C = \sqrt{10\,664}$$

$$= 103.3\,\Omega$$

and $\quad X_C = \dfrac{10^6}{2\pi fC} \quad$ (where C is the capacitance in microfarads)

$$\therefore \quad 103.3 = \frac{10^6}{2\pi \times 50 \times C}$$

$$\therefore \quad \frac{1}{103.3} = \frac{314 \times C}{10^6}$$

$$\therefore \quad C = \frac{10^6}{314 \times 103.3}$$

$$= 30.83\,\mu F$$

EXAMPLE 4   A coil of inductance 0.15 H and resistance 10 Ω is wired in series with a 60 μF capacitor to a 230 V 50 Hz supply (Fig. 6). Calculate the current which flows and the voltage drop across the capacitor.

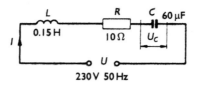

Fig. 6

Inductive reactance $X_L = 2\pi f L$

$$= 2\pi \times 50 \times 0.15$$

$$= 47.1\,\Omega$$

Capacitive reactance $X_C = \dfrac{10^6}{2\pi f C}$

$$= \dfrac{10^6}{314 \times 60}$$

$$= 53.1\,\Omega$$

Resultant reactance $X = X_C - X_L$ (take smaller from larger)

$$= 53.1 - 47.1$$

$$= 6.0\,\Omega$$

To find impedance,

$$Z^2 = R^2 + X^2$$

$$= 10^2 + 6.0^2$$

$$= 100 + 36$$

$$= 136$$

$\therefore \qquad Z = \sqrt{136}$

$$= 11.66\,\Omega$$

To find the current,

$$U = I \times Z$$

$\therefore \qquad 230 = I \times 11.66$

$\therefore \qquad I = \dfrac{230}{11.66}$

$$= 19.73\,\text{A}$$

Voltage across capacitor, $U_C = I \times X_C$

$$= 19.73 \times 53.1$$

$$= 1048\,\text{V}$$

EXERCISE 4

1. Complete the following table:

R	15	25	3.64				76.4	0.54
R^2				2250	18.7	40		

2. Complete the following table:

X	29.8		0.16			897		
X^2		0.46		0.9	0.16		54 637	0.036

3. A coil of wire has resistance of $8\,\Omega$ and inductance of 0.04 H. It is connected to a supply of 100 V at 50 Hz. Calculate the current which flows.

4. An inductor of inductance 0.075 H and resistance $12\,\Omega$ is connected to a 230 V supply at 50 Hz. Calculate the current which flows.

5. Complete the following table:

$R\,(\Omega)$	14.5		9.63	3.5	57.6		
$X\,(\Omega)$	22.8	74.6		34.7		49.6	
$Z\,(\Omega)$		159	18.4		4050	107	

6. A capacitor of $16\,\mu\text{F}$ and a resistor of $120\,\Omega$ are connected in series. Calculate the impedance of the circuit.

7. A resistor of $200\,\Omega$ and a capacitor of unknown value are connected to a 230 V supply at 50 Hz and a current of 0.85 A flows. Calculate the value of the capacitor in microfarads.

8. When a certain coil is connected to a 110 V d.c. supply, a current of 6.5 A flows. When the coil is connected to a 110 V 50 Hz a.c. supply, only 1.5 A flows. Calculate (a) the resistance of the coil, (b) its impedance, and (c) its reactance.

9. The inductor connected in series with a mercury vapour lamp has resistance of $2.4\,\Omega$ and impedance of $41\,\Omega$. Calculate the inductance of the inductor and the voltage drop across it when the total current in the circuit is 2.8 A. (Assume the supply frequency is 50 Hz.)

10. An inductor takes 8 A when connected to a d.c. supply at 230 V. If the inductor is connected to an a.c. supply at 230 V 50 Hz, the current is 4.8 A.

Calculate (a) the resistance, (b) the inductance, and (c) the impedance of the inductor.

11. What is the function of an inductor in an alternating-current circuit?

    When a d.c. supply at 230 V is applied to the ends of a certain inductor coil, the current in the coil is 20 A. If an a.c. supply at 230 V 50 Hz is applied to the coil, the current in the coil is 12.15 A. Calculate the impedance, reactance, inductance, and resistance of the coil.

    What would be the general effect on the current if the frequency of the a.c. supply were increased?

12. A coil having constant inductance of 0.12 H and resistance of 18 Ω is connected to an alternator which delivers 100 V a.c. at frequencies ranging from 28 to 55 Hz. Calculate the impedance of the coil when the frequency is 30, 35, 40, 45, and 50 Hz and plot a graph showing how the current through the coil varies according to the frequency.

13. The inductor in a discharge lighting circuit causes a voltage drop of 120 V when the current through it is 2.6 A. Determine the size in microfarads of a capacitor which will produce the same voltage drop at the same current value. (Neglect the resistance of the inductor. Assume the supply frequency is 50 Hz.)

14. A circuit is made up of an inductor, a resistor, and a capacitor all wired in series. When the circuit is connected to a 50 Hz a.c. supply, a current of 2.2 A flows. A voltmeter connected to each of the components in turn indicates 220 V across the inductor, 200 V across the resistor, and 180 V across the capacitor. Calculate the inductance of the inductor and the capacitance of the capacitor. At what frequency would these two components have the same reactance value? (Neglect the resistance of the inductor.)

15. What are meant by the following terms used in connection with alternating current: resistance, impedance, and reactance?

    A voltage of 230 V, at a frequency of 50 Hz, is applied to the ends of a circuit containing a resistor of 5 Ω, an inductor of 0.02 H, and a capacitor of 150 μF, all in series. Calculate the current in the circuit.

16. A coil of resistance 20 Ω and inductance 0.08 H is connected to a supply at 240 V 50 Hz. Calculate (a) the current in the circuit, (b) the value of a capacitor to be put in series with the coil so that the current shall be 12 A.   (CGLI)

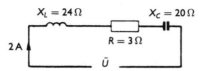

Fig. 7

17. For the circuit shown in Fig. 7, the voltage V is
    (a) 94 V **(b)** 14 V **(c)** 10 V **(d)** 0.043 V

18. An inductor has inductance 0.12 H and resistance 100 Ω. When it is connected to a 100 V supply at 150 Hz, the current through it is
    (a) 1.51 A **(b)** 0.47 A **(c)** 0.66 A **(d)** 0.211 A

IMPEDANCE TRIANGLES AND POWER TRIANGLES

For a right-angled triangle (Fig. 8), the theorem of Pythagoras states that

$$a^2 = b^2 + c^2$$

Fig. 8

As the relationship between impedance, resistance, and reactance in a series circuit is given by an equation of a similar form, $Z^2 = R^2 + X^2$, conditions in such circuits can conveniently be represented by right-angled triangles. In Fig. 9,

$$Z^2 = R^2 + X^2$$

where $X = X_L$ (Fig. 9(a))   or X_C (Fig. 9(b))

and ϕ = the *phase angle* of the circuit

$$\sin \phi = \frac{X}{Z} \qquad \cos \phi = \frac{R}{Z} \quad \text{and} \quad \tan \phi = \frac{X}{R}$$

Cos ϕ is the *power factor* of the circuit.

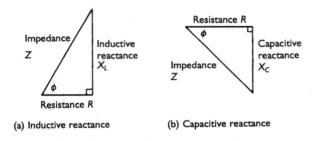

(a) Inductive reactance   (b) Capacitive reactance

Fig. 9

A right-angled triangle is also used to represent the apparent power in a circuit and its active and reactive components (Fig. 10).

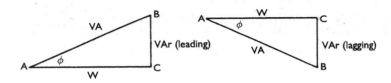

Fig. 10

AB is the product of voltage and current in the circuit (VA).
AC is the true power − the working component (W).
BC is the reactive or wattless component (VAr).

$$\frac{\text{VAr}}{\text{VA}} = \sin\phi$$

$\therefore$ $\quad$ VAr $= $ VA $\times \sin\phi$

$$\frac{\text{W}}{\text{VA}} = \cos\phi$$

$\therefore$ $\quad$ W $= $ VA $\cos\phi$

and $\cos\phi$ is the power factor (p.f.).

In power circuits, the following multiples of units are used:

kVA   kW   and   kVAr

EXAMPLE I Find Z in Fig. 11.

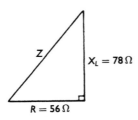

Fig. 11

$$Z^2 = R^2 + X_L{}^2$$
$$= 56^2 + 78^2$$
$$= 3135 + 6084$$
$$= 9219$$
$$\therefore \quad Z = \sqrt{9219}$$
$$= 96.02$$
$$= 96\,\Omega \text{ (correct to three significant figures)}$$

EXAMPLE 2 Find X_C in Fig. 12.

$$Z^2 = R^2 + X_C{}^2$$
$$125^2 = 67.6^2 + X_C{}^2$$
$$\therefore \quad X_C{}^2 = 125^2 - 67.6^2$$
$$= 15\,620 - 4570$$
$$= 11\,050$$
$$\therefore \quad X_C = \sqrt{11\,050} = 105.1$$
$$= 105\,\Omega$$

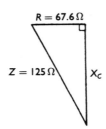

Fig. 12

Alternatively,

$$Z^2 = R^2 + X_C{}^2$$

$$125^2 = 67.6^2 + X_C{}^2$$

$$\therefore \quad X_C{}^2 = 125^2 - 67.6^2$$

$$= (125 + 67.6)(125 - 67.6)$$

$$= 192.6 \times 57.4$$

$$= 11\,050$$

$$\therefore \quad X_C = \sqrt{11\,050}$$

$$= 105\,\Omega$$

EXAMPLE 3 Find ϕ in Fig. 13.

$$\tan \phi = \frac{X_L}{R}$$

$$= \frac{15}{20} = 0.75$$

$$\therefore \quad \phi = 36° \ 52'$$

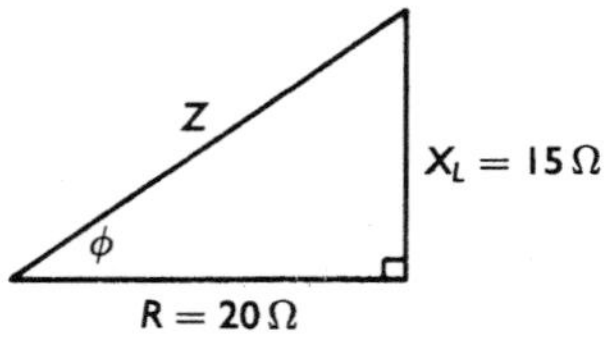

Fig. 13

EXAMPLE 4 Find X_C in Fig. 14.

$$\frac{X_C}{Z} = \sin \phi$$

$$\frac{X_C}{90} = \sin 48° = 0.7431$$

$$\therefore \quad X_C = 90 \times 0.7431$$

$$= 66.9 \text{ (to three significant figures)}$$

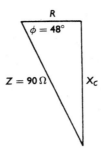

Fig. 14

EXAMPLE 5 Find the kVA and kVAr in Fig. 15.

$$\frac{kW}{kVA} = \cos\phi$$

$$\frac{15}{kVA} = \cos 42° = 0.7431$$

$$\therefore \quad \frac{kVA}{15} = \frac{1}{0.7431}$$

$$\therefore \quad kVA = \frac{15}{0.7431}$$

$$= 20.2$$

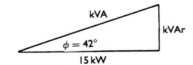

Fig. 15

$$\frac{kVAr}{kW} = \tan\phi$$

$$\therefore \quad \frac{kVAr}{15} = \tan 42° = 0.9004$$

$$\therefore \quad kVAr = 15 \times 0.9004$$

$$= 13.5$$

EXAMPLE 6 A coil of 0.2 H inductance and negligible resistance is connected in series with a 50 Ω resistor to the 230 V 50 Hz mains (Fig. 16). Calculate (a) the current which flows, (b) the power factor,

(c) the phase angle between the current and the applied voltage.

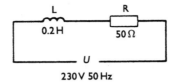

Fig. 16

Coil reactance $X_L = 2\pi f L$

$$= 2\pi \times 50 \times 0.2$$

$$= 314 \times 0.2$$

$$= 62.8\,\Omega$$

To find the impedance (Fig. 17),

$$Z^2 = R^2 + X_L{}^2$$

$$= 50^2 + 62.8^2$$

$$= 2500 + 3944$$

$$= 6444$$

$\therefore \qquad Z = \sqrt{6444}$

$$= 80.27\,\Omega$$

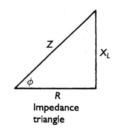

Impedance
triangle

Fig. 17

(a) To find the current,

$$U = I \times Z$$

$\therefore \qquad 230 = I \times 80.27$

$\therefore \qquad I = \dfrac{230}{80.27}$

$$= 2.86\,\text{A}$$

(b) Power factor $= \cos \phi = \dfrac{R}{Z}$

$$= \frac{50}{80.27}$$

$$= 0.623 \text{ lag}$$

(c) The phase angle is the angle whose cosine is 0.623,

$\therefore \qquad \phi = 51° \ 28'$

EXERCISE 5

1. Find Z in Fig. 18.

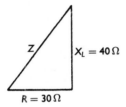

Fig. 18

2. Find Z in Fig. 19.

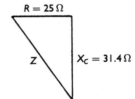

Fig. 19

3. Find R in Fig. 20.

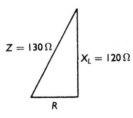

Fig. 20

4. Find X_C in Fig. 21.

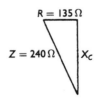

Fig. 21

5. Find R in Fig. 22.

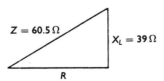

Fig. 22

6. Find Z in Fig. 23.

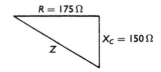

Fig. 23

7. Find R in Fig. 24.

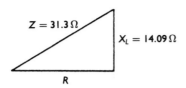

$Z = 31.3\,\Omega$
$X_L = 14.09\,\Omega$
R

Fig. 24

8. Find X_L in Fig. 25.

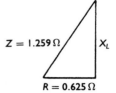

$Z = 1.259\,\Omega$
X_L
$R = 0.625\,\Omega$

Fig. 25

9. Find Z in Fig. 26.

Z
$X_C = 354\,\Omega$

Fig. 26

10. Find X_L in Fig. 27.

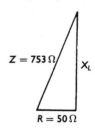

$Z = 753\,\Omega$
X_L
$R = 50\,\Omega$

Fig. 27

11. Find R in Fig. 28.

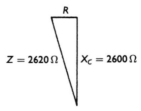

R
$Z = 2620\,\Omega$
$X_C = 2600\,\Omega$

Fig. 28

12. Consider the answers to questions 9 to 11 and then write down the approximate impedance of a coil which has resistance $32\,\Omega$ and reactance $500\,\Omega$.

13. Complete the following table:

Angle ϕ	30°	45°	60°	90°	52° 24′	26° 42′	83° 12′	5° 36′
sin ϕ								
cos ϕ								
tan ϕ								

14. Complete the following table:

Angle ϕ	33° 3′	75° 21′	17° 15′	64° 29′	27° 56′	41° 53′
sin ϕ						
cos ϕ						
tan ϕ						

15. Complete the following table:

Angle ϕ						
sin ϕ			0.91	0.6		0.9088
cos ϕ		0.9003		0.8		0.4754
tan ϕ	0.4000				1.2088	

16. Complete the following table:

Angle ϕ			38° 34′				
sin ϕ	0.9661						
cos ϕ		0.4341			0.8692	0.3020	0.318
tan ϕ			0.0950			3.15	

17. Find R and X_L in Fig. 29.   **18.** Find R and X_C in Fig. 30.

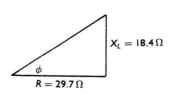

Fig. 29

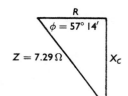

Fig. 30

19. Find ϕ in Fig. 31.   **20.** Calculate Z and X_L in Fig. 32.

Fig. 31

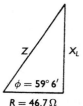

Fig. 32

21. Find W and VAr in Fig. 33.

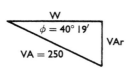

Fig. 33

22. Find ϕ and X_L in Fig. 34.

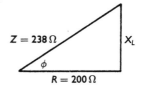

Fig. 34

23. Find ϕ in Fig. 35.

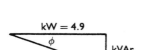

Fig. 35

24. Calculate R in Fig. 36.

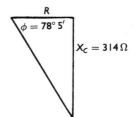

Fig. 36

25. Find OX in Fig. 37.

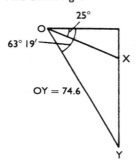

Fig. 37

26. Find OX in Fig. 38.

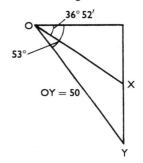

Fig. 38

27. Complete the following table and then plot a graph of power factor ($\cos \phi$) to a base of phase angle (ϕ):

Phase angle ϕ			65° 6′	60°			45° 40′	
Power factor $\cos \phi$	0.25	0.3			0.55	0.6		0.82

28. A coil has inductance 0.18 H and resistance 35 Ω. It is connected to a 100 V 50 Hz supply. Calculate (a) the impedance of the coil, (b) the current which flows, (c) the power factor, (d) the power absorbed by the coil.

29. Define the term 'power factor' and state how it affects cable size.

An inductor of resistance 8 Ω and of inductance 0.015 H is connected to an alternating-current supply at 230 V, single-phase, 50 Hz. Calculate (a) the current from the supply, (b) the power in the circuit, (c) the power factor.

30. A single-phase alternating-current supply at 230 V 50 Hz is applied to a series circuit consisting of an inductive coil of negligible resistance and a non-inductive resistance coil of 15 Ω. When a voltmeter is applied to the ends of each coil in turn, the potential differences are found to be 127.5 V across the inductive coil, 203 V across the resistance.

Calculate (a) the impedance of the circuit, (b) the inductance of the coil, (c) the current in the circuit, and (d) the power factor.   (CGLI)

31. On what factors do the resistance, reactance, and impedance of an alternating-current circuit depend, and how are these quantities related?

The current in a single-phase circuit lags behind the voltage by 60°. The power in the circuit is 3600 W and the voltage is 240 V. Calculate the value in ohms of the resistance, the reactance, and the impedance.   (CGLI)

Waveform and phasor representation of alternating currents and voltages

ALTERNATING E.M.F. AND CURRENT

The value and direction of the e.m.f. induced in a conductor rotating at constant speed in a uniform magnetic field, Fig. 39(a), vary according to the position of the conductor.

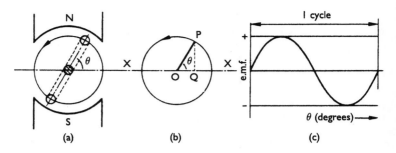

Fig. 39

The e.m.f. can be represented by the displacement QP of the point P above the axis XOX, Fig. 39(b). OP is a line which is rotating about the point O at the same speed as the conductor is rotating in the magnetic field. The length of OP represents the maximum value of the induced voltage. OP is called a *phasor*.

A graph, Fig. 39(c), of the displacement of the point P plotted against the angle θ (the angle through which the conductor has moved from the position of zero induced e.m.f.) is called a *sine wave*, since the PQ is proportional to the sine angle θ. One complete revolution of OP is called a *cycle*.

EXAMPLE I An alternating voltage has a maximum value of 200 V. Assuming that it is sinusoidal in nature (i.e. it varies according to a sine wave), plot a graph to show the variations in this voltage over a complete cycle.

Method (Fig. 40)   Choose a reasonable scale for OP; for instance, 10 mm ≡ 100 V.

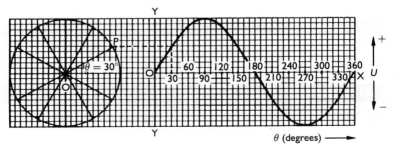

Fig. 40

Draw a circle of radius 20 mm at the left-hand side of a piece of graph paper to represent the rotation of OP.

One complete revolution of OP sweeps out $360°$. Divide the circle into any number of equal portions, say 12. Each portion will then cover $30°$.

Construct the axes of the graph, drawing the horizontal axis OX (the *x*-axis) on a line through the centre of the circle. This *x*-axis should now be marked off in steps of $30°$ up to $360°$. If desired, perpendicular lines can be drawn through these points. Such lines are called *ordinates*.

The points on the graph are obtained by projecting from the various positions of P to the ordinate corresponding to the angle θ at that position.

Remember that when $\theta = 0°$ and $180°$ the generated e.m.f. is zero, and when $\theta = 90°$ and $270°$ the generated e.m.f. has its maximum value.

EXAMPLE 2   Two alternating voltages act in a circuit. One (A) has an r.m.s. value of 90 V and the other (B) has an r.m.s. value of 40 V, and A leads B by $80°$. Assuming that both voltages are sinusoidal, plot graphs to show their variations over a complete cycle. By adding their instantaneous values together, derive a graph of the resultant voltage. Give the r.m.s. value of this resultant.

First find the maximum values of the voltages given:

$$U_{\text{r.m.s.}} = 0.707 \times U_{\text{max.}}$$

$$\therefore \qquad 90 = 0.707 \times U_{\text{max.}}$$

$$\therefore \qquad U_{\text{max.}} = \frac{90}{0.707}$$

$$= 127 \, \text{V}$$

Similarly, if

$$U_{\text{r.m.s.}} = 40$$

$$U_{\text{max.}} = \frac{40}{0.707}$$

$$= 56.6 \, \text{V}$$

Choose a suitable scale, say 20 mm ≡ 100 V. Draw two circles with the same centre, one having a radius of 25.4 mm (127 V), the other a radius of 11.32 mm (56.6 V).

Draw phasors to represent the voltages: OA horizontal and OB, which represents the lower voltage, lagging 80° behind OA (anticlockwise rotation is always used) – see Fig. 41.

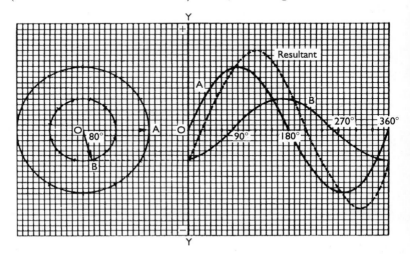

Fig. 41

Mark off the circumference of the larger circle in steps of 30°, using OA as the reference line.

Mark off the smaller circle in steps of 30°, using OB as the reference line.

Set off the axes of the graph alongside as in the previous example.

Plot the sine wave of voltage A as before.

Plot the sine wave of voltage B in exactly the same way, projecting the first point from B to the y-axis YOY and from each succeeding 30° point to the appropriate 30° point on the horizontal axis of the graph.

Points on the resultant graph are plotted by combining the ordinates of A and B at each 30° point. If the graphs lie on the same side of the x-axis, the ordinates are added. If the graphs lie on opposite sides of the axis, the smaller is subtracted from the larger (measurements upwards from the x-axis are positive, measurements downwards are negative).

The resultant curve is shown by the dotted line in Fig. 41 and its maximum value is approximately 150 V.

Its r.m.s. value is

$$0.707 \times 150 = 106\,\text{V}$$

EXAMPLE 3   A current of 15 A flows from the 230 V mains at a power factor of 0.76 lagging. Assuming that both current and voltage are sinusoidal, plot graphs to represent them over one cycle. Plot also on the same axes a graph showing the variation in power supplied over one cycle.

The procedure for plotting the current and voltage sine waves is the same as that adopted in the previous example.

The phase angle between current and voltage is found from the power factor as follows:

$$\text{power factor} = \cos\phi$$

where ϕ is the angle of phase difference

$$\cos\phi = 0.76$$

$$\therefore \qquad \phi = 40°\,32'$$

$$U_{\text{max.}} = \frac{230}{0.707}$$

$$= 325.3\,\text{V}$$

$$I_{\text{max.}} = \frac{15}{0.707}$$

$$= 21.21\,\text{A}$$

Scales of 20 mm ≡ 200 V and 20 mm ≡ 20 A will be suitable.

To obtain the graph of the power supplied, the ordinates of current and voltage are multiplied together (Fig. 42). It is convenient to do this every 30° as before.

Remember the rules for multiplying positive and negative numbers.

Where the resulting graph is negative, additional points are helpful in obtaining a smooth curve.

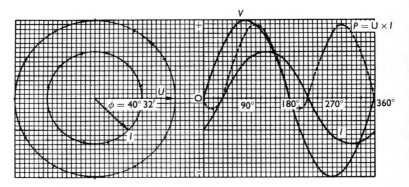

Fig. 42

That portion of the power curve lying above the *x*-axis represents the power supplied to the circuit. That portion lying below the *x*-axis represents the power returned to the mains from the circuit.

1. Plot a sine wave, over one complete cycle, of an alternating voltage having a maximum value of 325 V. Determine the r.m.s. value of this voltage.

2. An alternating current has the following values taken at intervals of 30° over one half cycle:

Angle ϕ	0	30°	60°	90°	120°	150°	180°
Current (A)	0	10.5	17.5	19.7	15.0	11.5	0

Determine the average and r.m.s. values of this current.

3. Plot a graph over one complete cycle of a sinusoidal alternating voltage having an r.m.s. value of 200 V.

4. Two sinusoidal voltages act in a circuit. Their r.m.s. values are 110 V and 80 V and they are out of phase by 75°, the lower voltage lagging. Plot sine waves on the same axes to represent these voltages. Plot a graph of the resultant voltage by adding together the ordinates of the two waves. Give the r.m.s. value of the resultant voltage and state approximately the phase angle between this resultant and the lower voltage.

5. Two alternating currents are led into the same conductor. They are sinusoidal and have r.m.s. values of 4 A and 1 A. The smaller current leads by 120°. Plot out the sine waves of these two currents and add the

ordinates to obtain the sine wave of the resultant current. Calculate the r.m.s. value of the resultant.

6. The current taken by an immersion heater from the 250 V a.c. mains is 12.5 A. Current and voltage are in phase and are sinusoidal. Plot graphs on the same axes to show the variations in current and voltage over one complete cycle.

7. A 10 μF capacitor is connected to a 240 V supply at 50 Hz. The current leads the voltage by 90°, and both may be assumed to be sinusoidal. Plot the sine waves of the current and voltage over one complete cycle.

8. A fluorescent lamp takes a current of 1.2 A from a 230 V supply at a power factor of 0.47. Assuming that both current and voltage are sinusoidal, plot graphs to show how they vary over a complete cycle.

9. The current in a circuit is 25 A and the supply voltage is 220 V. The power factor is 0.6 lagging. Plot sine waves to represent current and voltage over one cycle. Multiply the instantaneous values of current and voltage together to obtain a graph representing the power in the circuit.

10. An inductor of 0.1 H is connected to a 100 V supply at 50 Hz. Neglecting the resistance of the winding, calculate the current which flows. Plot sine waves to represent the current and voltage, assuming that the voltage leads the current by 90°. Multiply the ordinates of the waves together to obtain a graph representing the power supplied to the circuit.

PHASORS

Conditions in alternating-current circuits can be represented by means of phasor diagrams.

In Fig. 43, U is a voltage and I is a current, ϕ is the angle of phase difference, and $\cos \phi$ is the power factor.

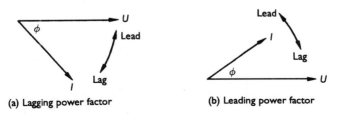

(a) Lagging power factor   (b) Leading power factor

Fig. 43

EXAMPLE I   The current in a circuit is 5 A, the supply voltage is 230 V, and the power factor is 0.8 lagging. Represent these conditions by means of a phasor diagram drawn to scale. Choose a suitable scale.

$$\text{Power factor} = 0.8$$
$$= \cos\phi$$
$$\cos\phi = 0.8$$
$$\therefore \qquad \phi = 36° 52' \quad \text{(see Fig. 44)}$$

U = 230 V
φ = 36° 52'
I = 5 A

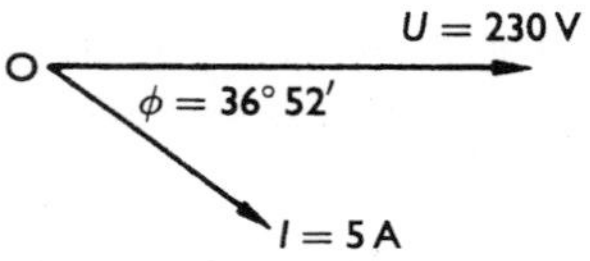

Fig. 44

Normally the r.m.s. values are used when drawing phasor diagrams.

Note that the most accurate construction is obtained by setting off two lines at the required angle and then marking the lines to the appropriate lengths from the point of intersection with compasses which have been set to the respective measurement.

EXAMPLE 2   A resistor and a capacitor are wired in series to an a.c. supply (Fig. 45). When a voltmeter is connected across the resistor it reads 150 V. When it is connected to the capacitor terminals it indicates 200 V. Draw the phasor diagram for this circuit to scale and thus determine the supply voltage.

As the value of current is not given, it will not be possible to draw its phasor to scale.

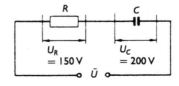

Fig. 45

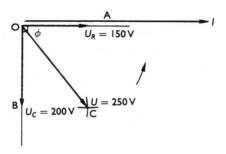

Fig. 46

The current is the same throughout a series circuit and so the current phasor is used as a reference.

Draw OI any length to represent the current (Fig. 46).

From point O, draw thin lines parallel to and at right angles to OI (capacitor voltage *lags* behind the current).

Choose a suitable scale and use compasses set to the required measurement to mark off OA = U_R, the resistor voltage drop – in phase with the current – and OB = U_C, the capacitor voltage drop.

With centre A and compasses set to length OB, strike an arc. With centre B and compasses set to OA, strike another arc. These arcs intersect at point C.

OC is the resultant voltage, which is equal to the supply voltage.

By measurement of OC, the supply voltage is found to be 250 V.

EXAMPLE 3 An inductor takes a current of 5 A from a 230 V supply at a power factor of 0.4 lagging. Construct the phasor diagram accurately to scale and estimate from the diagram the resistance and reactance of the coil.

As already explained, although resistance and reactance cannot be separated, it is convenient to draw them apart in an equivalent circuit diagram (Fig. 47). The total voltage drop – in this case the supply voltage – will then be seen to be made up of a resistance voltage drop and a reactance voltage drop.

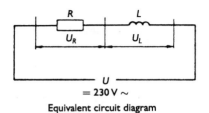

Equivalent circuit diagram

Fig. 47

Since, again, we are considering a series circuit in which the current is the same throughout, it is not necessary to draw the current phasor to scale.

Power factor $= \cos \phi$

where ϕ is the angle of phase difference between current and supply voltage

and $\cos \phi = 0.4$

$\therefore \qquad \phi = 66° 25'$

Draw OI any length to represent the current (Fig. 48).

Choose a suitable scale and set off OC at $66° 25'$ from OI and of length to represent the supply voltage.

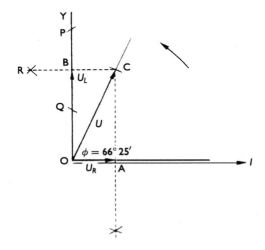

Fig. 48

Draw OY at right angles to the current phasor and from C draw perpendiculars to cut the current phasor at A and OY at B. The perpendiculars are constructed as follows:

(i) Set the compasses to any radius and with centre C draw arcs which cut OY at P and Q.

(ii) With the compasses again set to any radius and with centres P and Q, strike two more arcs to cut in R. CR is then perpendicular to OY.

A similar method is employed in drawing CA.

By measurement,

$$U_R = 93\,\text{V}$$

$$U_L = 209\,\text{V}$$

Now $\quad U_R = I \times R$

$\therefore \qquad 93 = 5 \times R$

$\therefore \qquad R = \dfrac{93}{5}$

$\qquad\qquad = 18.5\,\Omega$

and $\quad U_L = I \times X_L$ (X_L is the inductive reactance)

$\therefore \qquad 209 = 5 \times X_L$

$\therefore \qquad X_L = \dfrac{209}{5}$

$\qquad\qquad = 41.8\,\Omega$

EXAMPLE 4   An appliance takes a single-phase current of 32 A at 0.6 p.f. lagging from a 250 V a.c. supply. A capacitor which takes 8.9 A is wired in parallel with this appliance (Fig. 49). Determine graphically the total supply current.

As this is a parallel circuit, the voltage is common to both branches and is thus used as the reference phasor. It need not be drawn to scale.

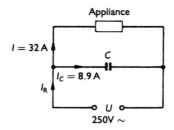

Fig. 49

Choose a suitable scale.

$$\text{p.f.} = \cos \phi = 0.6$$

$$\therefore \qquad \phi = 53° \, 8'$$

Draw the voltage phasor (Fig. 50) and set off the appliance-current phasor at 53° 8' lagging (OA).

The capacitor current, 8.9 A, leads on the voltage by 90° and is drawn next (OB).

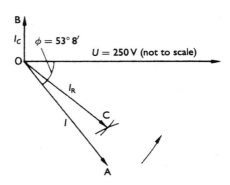

Fig. 50

The resultant of these two phasors is found as follows:

(i) With compasses set to OA and centre B, strike an arc.

(ii) With centre A and compasses set to OB, strike another arc cutting the first in C.

OC is the resultant current. By measurement of OC, the resultant current is 25.5 A.

EXAMPLE 5 A consumer's load is 15 kVA single-phase a.c. at 0.8 power factor lagging. By graphical construction, estimate the active and reactive components of this load.

$$\text{p.f.} = \cos\phi = 0.8$$

$$\therefore \qquad \phi = 36° \, 52'$$

Choose a suitable scale.

Draw a thin horizontal line OX (Fig. 51). Set off OA to represent 15 kVA at an angle of $36° \, 52'$ from OX.

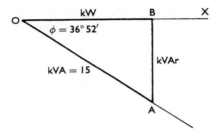

Fig. 51

From A, draw a perpendicular to cut line OX at B. OB is then the working or active component and AB is the reactive or wattless component.

By measurement of OB the true power is 12 kW, and by measurement of AB the wattless component is 9 kVAr.

EXERCISE 7

1. An a.c. circuit takes a current of 15 A at a power factor of 0.75 lagging from the 230 V mains. Construct, to scale, the phasor diagram for this circuit.

2. A power-factor-improvement capacitor takes a current of 1.6 A from a 230 V supply. Draw the phasor diagram to scale.

3. A single-phase a.c. motor takes a current of 2.75 A at a power factor of 0.18 lagging when it is running on no load. On full load it takes 4.3 A at a power factor of 0.48 lagging. The supply voltage is in each case 230 V. Draw a phasor diagram to represent the no-load and full-load circuit conditions.

4. A mercury-vapour-lamp circuit takes a current of 2.8 A at a power factor of 0.45 lagging if it is used without its p.f. improvement capacitor.

When the p.f. improvement capacitor is connected, the current falls to 1.8 A at 0.7 p.f. lagging. Construct the phasor diagram to scale.

5. A capacitor is wired in series with a resistor to an a.c. supply. When a voltmeter is connected to the capacitor terminals it indicates 180 V. When it is connected across the resistor it reads 170 V. Construct the phasor diagram for this circuit accurately to scale and from it determine the supply voltage.

6. An inductor has resistance $10\,\Omega$ and when it is connected to a 240 V a.c. supply a current of 12 A flows. Draw the phasor diagram to scale.

7. A contactor coil takes a current of 0.085 A from a 250 V supply at a power factor of 0.35 lagging. Draw the phasor diagram accurately to scale and use it to determine the resistance and reactance of the coil.

8. A single-phase transformer supplies 10 kVA at 0.7 p.f. lagging. Determine by graphical construction the active and reactive components of this load.

9. The true power input to a single-phase motor is 1150 W and the power factor is 0.54 lagging. Determine graphically the apparent power input to the machine.

10. A fluorescent-lamp circuit takes a current of 1.2 A at 0.65 p.f. lagging from the 230 V a.c. mains. Determine graphically the true power input to the circuit.

11. A single-phase motor takes 8.5 A from a 230 V supply at 0.4 p.f. lagging. A capacitor which takes 4 A is connected in parallel with the motor. From a phasor diagram drawn accurately to scale, determine the resultant supply current.

12. A discharge lighting fitting takes a current of 5.2 A at 0.46 p.f. lagging when it is used without its power-factor-improvement capacitor. When this capacitor is connected the current falls to 3.2 A, the supply voltage remaining constant at 240 V. Draw the phasor diagram to represent the conditions with and without the capacitor and from it determine the current taken by the capacitor. (Remember that the working component of the supply current is constant.)

13. A series circuit is made up of a resistor, an inductor of negligible resistance, and a capacitor. The circuit is connected to a source of alternating current, and a voltmeter connected to the terminals of each component in turn indicates 180 V, 225 V, and 146 V respectively. Construct the phasor diagram for this circuit accurately to scale and hence determine the supply voltage.

PARALLEL CIRCUITS INVOLVING RESISTANCE, INDUCTANCE AND CAPACITANCE

Consider a circuit consisting of inductance and capacitance in parallel (Fig. 52).

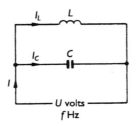

Fig. 52

L is a pure inductance (henry).

C is a pure capacitance (microfarad).

The same voltage is applied to each branch of the circuit.

The current through the inductance is $I_L = U/X_L$, where $X_L = 2\pi f L$. This current lags the voltage by $90°$.

The capacitor current is $I_C = U/X_C$, where $X_C = 10^6/2\pi f C$. This current leads the voltage by $90°$.

The phasor diagram is usually drawn as in Fig. 53.

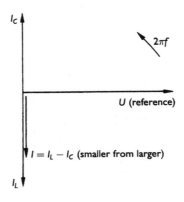

Fig. 53

EXAMPLE 1   If in Fig. 52 the reactance of the inductor is $100\,\Omega$ and that of the capacitor is $150\,\Omega$, determine the resultant current when the supply is at $100\,\text{V}$.

$$\text{Inductor current } I_L = \frac{U}{X_L}$$

$$= \frac{100}{100} = 1\,\text{A}$$

lagging the supply voltage by $90°$.

$$\text{Capacitor current } I_C = \frac{U}{X_C}$$

$$= \frac{100}{150} = 0.667\,\text{A}$$

leading the supply voltage by $90°$.

By reference to the phasor diagram in Fig. 53, the resultant supply current is seen to be

$$I = 1 - 0.667$$

$$= 0.33\,\text{A}$$

lagging the supply voltage by $90°$.

EXAMPLE 2   Determine the resultant current when a capacitor of $22\,\mu\text{F}$ is connected in parallel with a resistor of $75\,\Omega$ to a $100\,\text{V}$ supply at $150\,\text{Hz}$ (Fig. 54). Draw the phasor diagram and determine the phase relationship between the resultant current and the supply voltage.

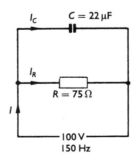

Fig. 54

$$X_C = \frac{10^6}{2\pi f C}$$

$$= \frac{10^6}{2\pi \times 150 \times 22}$$

$$= 48.23\,\Omega$$

$$I_C = \frac{U}{X_C}$$

$$= \frac{100}{48.23}$$

$$= 2.073\,\text{A}$$

$$I_R = \frac{100}{75}$$

$$= 1.33\,\text{A}$$

The phasor diagram is shown in Fig. 55.

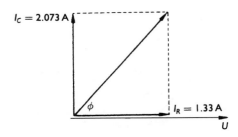

Fig. 55

From the phasor diagram it is seen that the supply current is given by

$$I = \sqrt{(I_R{}^2 + I_C{}^2)}$$

$$= \sqrt{(1.33^2 + 2.073^2)}$$

$$= 2.46\,\text{A}$$

The phase angle ϕ is given by

$$\tan\phi = \frac{2.073}{1.33} \quad \left(\frac{\text{opposite side}}{\text{adjacent side}}\right)$$

$$= 1.558$$

$$\therefore \qquad \phi = 57.3°$$

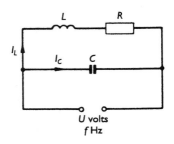

Fig. 56

Thus the resultant current is 2.46 A *leading* by 57.3° on the supply voltage.

Usually the inductive branch possesses resistance (Fig. 56). I_L then lags the voltage U by some angle less than 90° depending on the values of L and R. (L and R form a series circuit.)

The phasor diagram is as shown in Fig. 57.

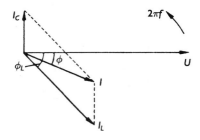

Fig. 57

ϕ_L is the phase angle of the inductive branch, and $\tan \phi_L = X_L/R$.

The supply current is the phasor sum of I_C and I_L, which is found by completing the parallelogram (Fig. 57). ϕ is the phase angle between the supply voltage and current.

EXAMPLE 3   A coil has resistance 25 Ω and inductive reactance 20 Ω. It is connected in parallel with a capacitor of reactance 40 Ω to a 230 V a.c. supply. Determine the supply current and the overall power factor.

The coil impedance, Z_L, is given by

$$Z_L{}^2 = R^2 + X_L{}^2$$
$$= 25^2 + 20^2$$
$$\therefore \quad Z_L = \sqrt{(25^2 + 20^2)}$$
$$= 32.02\,\Omega$$

Coil current $I_L = \dfrac{U}{Z_L}$
$$= \dfrac{230}{32.02}$$
$$= 7.183\,\text{A}$$
$$= 7.2\,\text{A}$$

The coil phase angle is found from

$$\tan\phi_L = \dfrac{X_L}{R}$$
$$\therefore \quad \tan\phi_L = \dfrac{20}{25}$$
$$= 0.8$$
$$\therefore \quad \phi_L = 38°\ 39' \quad \text{(from tables)}$$

Capacitor current $I_C = \dfrac{U}{X_C}$
$$= \dfrac{230}{40}$$
$$= 5.75\,\text{A}$$

The supply current I_S may be determined:

(a) graphically by constructing the phasor diagram to scale (Fig. 58). By measurement, $I_S = 5.75\,\text{A}$ and $\phi = 13°$, hence $\cos\phi = 0.97$ leading.

(b) by calculation as follows:

The horizontal component of the coil current is

$$I_L \cos\phi_L = 7.2\cos 38°\ 39'$$
$$= 7.2 \times 0.7810$$
$$= 5.623$$

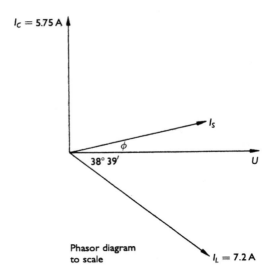

Fig. 58

$$\frac{\text{Horizontal component}}{\text{of capacitor current}} = 0$$

The total horizontal component is

$$X = 5.623 + 0$$

$$= 5.623$$

$$\frac{\text{Vertical component}}{\text{of capacitor current}} = 5.75\,\text{A}$$

The vertical component of the coil current is

$$-I_L \sin \phi_L = -7.2 \sin 38° 39'$$

$$= -7.2 \times 0.6246$$

$$= -4.5 \quad \text{(the minus sign because this}$$
$$\text{component acts downwards)}$$

The total vertical component is

$$Y = 5.75 - 4.5$$

$$= 1.25$$

The resultant current is

$$I_S = \sqrt{(X^2 + Y^2)}$$
$$= \sqrt{(5.623^2 + 1.25^2)}$$
$$= 5.76 \, \text{A}$$

The phase angle between the supply current and the voltage is given by

$$\tan \phi = \frac{Y}{X}$$
$$= \frac{1.25}{5.623}$$
$$= 0.2223$$
$$\therefore \qquad \phi = 12° \, 30'$$

and the overall power factor is

$$\cos \phi = \cos 12° \, 30'$$
$$= 0.976 \, \text{leading}$$

EXERCISE 8

1. Determine the current I in Fig. 59 and state whether it leads or lags the voltage U.

2. Determine the resultant current I and its phase relationship with the supply voltage U in Fig. 60. What is the power factor of the circuit?

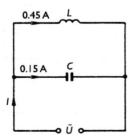

Fig. 59   **Fig. 60**

3. A capacitor of 15 μF is connected in parallel with a coil of inductance 0.3 H and negligible resistance to a sinusoidal supply of 240 V 50 Hz. Calculate the resultant current and state whether the phase angle is a leading or lagging one.

4. Calculate the resulting supply current and the overall power factor when a resistor of 100 Ω is connected in parallel with the circuit of question 3.

5. A coil of reactance 30 Ω and resistance 40 Ω is connected in parallel with a capacitor of reactance 200 Ω, and the circuit is supplied at 200 V. Calculate the resultant current and power factor.
Check the results by constructing the phasor diagram accurately to scale.

6. A coil has resistance 150 Ω and inductance 0.478 H. Calculate the value of a capacitor which when connected in parallel with this coil to a 50 Hz supply will cause the resultant supply current to be in phase with the voltage.

7. An inductor coil of resistance 50 Ω takes a current of 1 A when connected in series with a capacitor of 31.8 μF to a 240 V 50 Hz supply. Calculate the resultant supply current when the capacitor is connected in parallel with the coil to the same supply.

8. The resultant current I in Fig. 61 is
(a) 0.585 A   (b) 0.085 A   (c) 11.2 A   (d) 171 A

9. The resultant current I in Fig. 62 is
(a) 4 A   (b) 8.5 A   (c) 2.92 A   (d) 9.22 A

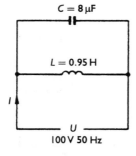

Fig. 61

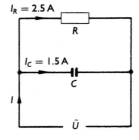

Fig. 62

Alternating-current motors

For a *single-phase* circuit:

> power (watts) = voltage × current × power factor
>
> $P = U \times I \times \text{p.f.}$

For a *three-phase* circuit:

> power (watts) = $\sqrt{3} \times$ line voltage × line current × power factor
>
> $P = \sqrt{3} \times U_L \times I_L \times \text{p.f.}$

EXAMPLE 1 Calculate the current taken by a 2 kW 230 V single-phase motor working at full load with an efficiency of 70% and power factor 0.6.

Output = 2 kW

$$\text{Input} = 2 \times 1000 \times \frac{100}{70}$$

$$= 2857\,\text{W}$$

$$P = U \times I \times \text{p.f.}$$

∴ $\quad 2857 = 230 \times I \times 0.6$

∴ $\quad I = \dfrac{2857}{230 \times 0.6}$

$$= 20.7\,\text{A}$$

EXAMPLE 2 The following results were recorded during a test on a single-phase a.c. motor:

Mechanical output = 1.2 kW

Supply volts = 230 V

Line current = 15 A

Power input = 1560 W

Calculate the efficiency of the motor and the power factor.

$$\text{Output power} = 1.2\,\text{kW}$$

$$= 1200\,\text{W}$$

$$\text{Input power} = 1560\,\text{W}$$

$$\text{Efficiency} = \frac{\text{output power}}{\text{input power}}$$

$$= \frac{1200}{1560}$$

$$= 0.769$$

$$\text{Percentage efficiency} = 0.769 \times 100$$

$$= 76.9\%$$

$$P = U \times I \times \text{p.f.}$$

$$\therefore \quad 1560 = 230 \times 15 \times \text{p.f.}$$

$$\therefore \quad \text{p.f.} = \frac{1560}{230 \times 15}$$

$$= 0.45$$

EXAMPLE 3 A load of 120 kg is raised through a vertical distance of 10 m in 45 s by a conveyor. The efficiency of the conveyor gear is 30% and that of the single-phase 230 V driving motor is 80%. The power factor of the motor is 0.8. Calculate the current taken by the motor.

The force required to raise a mass or load of 1 kg against the effect of gravity is 9.81 N.

$$\text{Work done on the load} = \text{distance} \times \text{force}$$

$$= 10\,\text{m} \times 120\,\text{kg} \times 9.81\,\text{N}$$

$$= 11\,772\,\text{N m}$$

and

$$1\,\text{N m} = 1\,\text{J}$$

Thus the work done on the load is 11 772 J.

$$\text{Work done per second} = \frac{11\,772}{45}\,\text{J/s}$$

$$= 261.6\,\text{J/s} \quad \text{or} \quad 261.6\,\text{W}$$

which is the power output from the conveyor.

The input to the conveyor which is the output of the driving motor is

$$261.6 \times \frac{100}{30} = 872 \, \text{W}$$

The input to the driving motor is

$$872 \times \frac{100}{80} = 1090 \, \text{W}$$

and the current taken by the motor is

$$\frac{1090}{230 \times 0.8} = 5.92 \, \text{A}$$

EXERCISE 9

1. Calculate the full-load current of each of the motors to which the following particulars refer:

	Power output (kW)	Phase	Voltage	Efficiency (%)	Power factor
(a)	5	1	230	70	0.7
(b)	3	1	250	68	0.5
(c)	15	3	400	75	0.8
(d)	6	1	230	72	0.55
(e)	30	3	400	78	0.7
(f)	0.5	1	230	60	0.45
(g)	8	3	400	65	0.85
(h)	25	3	440	74	0.75

2. You are required to record the input to a single-phase a.c. motor in kW and in kVA. Make a connection diagram showing the instruments you would use.

   A 30 kW single-phase motor delivers full-load output at 0.75 power factor. If the input is 47.6 kVA, calculate the efficiency of the motor.

3. A single-phase motor develops 15 kW. The input to the motor is recorded by instruments with readings as follows: 230 V, 100 A, and 17 590 W. Calculate the efficiency of the motor and its power factor. Draw a diagram of the connections of the instruments. Account for the energy lost in the motor.

4. Make a diagram showing the connections of a voltmeter, an ammeter, and a wattmeter, in a single-phase a.c. circuit supplying power to a motor.

The following values were recorded in a load test of a single-phase motor. Calculate the efficiency of the motor and its power factor:

Voltmeter reading 230 V

Ammeter reading 75 A

Wattmeter reading 13 kW

Mechanical output 10 kW

5. **(a)** What is power factor?

(b) Why is a.c. plant rated in kVA? Illustrate your answer by comparing the load on circuit cables of (i) a 10 kW d.c. motor and (ii) a 10 kW single-phase a.c. motor at the same supply, operating at a power factor of 0.7.

6. A single-phase motor drives a pump which raises 500 kg of water per minute to the top of a building 12 m high. The combined efficiency of the pump and motor is 52%, the supply voltage is 230 V, and the power factor is 0.45. Calculate the supply current.

7. The output of a motor is 75 kW; the input is 100 kW. The efficiency of the motor is

(a) 13.3%   **(b)** 7.5%   **(c)** 1.33%   **(d)** 75%

8. The efficiency of a motor is 80%. The input power when its output is 24 kW is

(a) 30 kW   **(b)** 19.2 kW   **(c)** 192 kW   **(d)** 300 kW

Power-factor improvement

EXAMPLE I   A consumer takes a load of 50 kVA at 0.7 power factor lagging. Calculate (a) the active and reactive components of this load, (b) the leading kVAr to be taken by a capacitor in order to improve the power factor to 0.9 lagging.

In Fig. 63, OA represents the true power (kW) and remains constant.

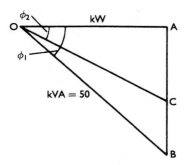

Fig. 63

OB represents the apparent power (kVA) before p.f. improvement.

AB represents the reactive power (kVAr) before p.f. improvement.

AC represents the reactive power (kVAr) after p.f. improvement.

$$\text{p.f. (before improvement)} = \cos \phi_1 = 0.7$$

$$\therefore \qquad \phi_1 = 45° \, 34'$$

$$\text{p.f. (after improvement)} = \cos \phi_2 = 0.9$$

$$\therefore \qquad \phi_2 = 25° \, 50'$$

(a) $\qquad \dfrac{OA}{OB} = \cos \phi_1$

$$\dfrac{OA}{50} = \cos 45° \, 34'$$

$$\therefore \qquad OA = 50 \times 0.7$$

$$= 35$$

i.e. active component $= 35 \, \text{kW}$

$$\dfrac{AB}{OB} = \sin \phi_1$$

$$\dfrac{AB}{50} = \sin 45° \, 34'$$

$$\therefore \qquad AB = 50 \times 0.7141$$

$$= 35.7$$

i.e. reactive component $= 35.7 \, \text{kVAr}$

(b) Reactive kVAr, leading, to be supplied by capacitor is

$$BC = AB - AC$$

Now $\dfrac{AB}{AO} = \tan \phi_1$ and $\dfrac{AC}{AO} = \tan \phi_2$

$\therefore$ $\dfrac{AB}{35} = \tan 45° \, 34'$ $\therefore$ $\dfrac{AC}{35} = \tan 25° \, 50'$

$\therefore$ $AB = 35 \times 1.02$ $\therefore$ $AC = 35 \times 0.4841$

thus $AB - AC = 35 \times 1.02 - 35 \times 0.4841$

$$= 35 \, (1.02 - 0.4841)$$

$$= 35 \times 0.5359$$

$$= 18.756$$

$\therefore$ leading kVAr to be taken by capacitor $= 18.8 \, \text{kVAr}$

EXAMPLE 2 A test on a 80 W fluorescent-lamp circuit (Fig. 64) yielded the following results:

without p.f. improvement capacitor:

volts 232
amperes 1.13
watts 122

with p.f. improvement capacitor:

volts 232
amperes 0.68
watts 122

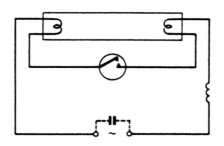

Fig. 64

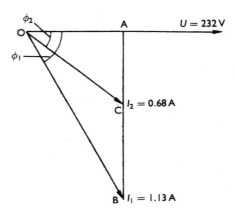

Fig. 65

The supply was taken from the 50 Hz mains. Calculate the value of the p.f. improvement capacitor in microfarads.

In Fig. 65,

U is the voltage phasor;

OB represents the current I_1 without the capacitor;

OC represents the current I_2 with the capacitor in circuit;

OA is the in-phase component of current – the same in each case;

AB is the wattless component of I_1;

AC is the wattless component of I_2.

In a single-phase a.c. circuit,

$$P = U \times I \times \text{p.f.}$$

$$= U \times I \times \cos \phi$$

$$\therefore \quad \cos \phi = \frac{P}{U \times I}$$

$$\cos \phi_1 = \frac{P}{U \times I_1}$$

$$= \frac{122}{232 \times 1.13} = 0.4654$$

$$\therefore \quad \phi_1 = 62° \ 16'$$

$$\cos \phi_2 = \frac{P}{V \times I_2}$$

$$= \frac{122}{232 \times 0.68} = 0.7734$$

Now CB represents the capacitor current and

$$CB = AB - AC$$

$$\frac{AB}{OB} = \sin \phi_1$$

$$\therefore \qquad \frac{AB}{1.13} = \sin 62° \ 16'$$

$$\therefore \qquad AB = 1.13 \times 0.8851$$

$$= 1.00$$

$$\frac{AC}{OC} = \sin \phi_2$$

$$\therefore \qquad \frac{AC}{0.68} = \sin 39° \ 20'$$

$$\therefore \qquad AC = 0.68 \times 0.6338$$

$$= 0.431$$

$$\therefore \quad AB - AC = 1.00 - 0.431$$

$$= 0.569$$

$$\therefore \quad \text{capacitor current} = 0.569 \, \text{A}$$

In the capacitor circuit,

$$U = I \times X_C$$

$$\therefore \qquad 232 = 0.569 \times X_C$$

$$\therefore \qquad X_C = \frac{232}{0.569}$$

$$= 407.8 \, \Omega$$

$$X_C = \frac{10^6}{2\pi f C}$$

$$\therefore \qquad 407.8 = \frac{10^6}{314 \times C}$$

$$\therefore \quad C = \frac{10^6}{407.8 \times 314}$$

$$= 7.809$$

$$= 7.81 \, \mu F$$

1. The nameplate of a single-phase transformer gives its rating as 5 kVA at 230 V. What is the full-load current that this transformer can supply and what is its power output when the load power factor is (a) 0.8, (b) 0.6?

2. **(a)** What is meant by power factor?

   (b) The installation in a factory carries the following loads:
   lighting 50 kW, heating 30 kW, and power 44 760 W.
   Assuming that the lighting and heating loads are non-inductive, and the power has an overall efficiency of 87% at a power factor of 0.7 lagging, calculate (i) the total loading in kW, (ii) the kVA demand at full load. (CGLI)

3. The current taken by a 230 V 50 Hz, single-phase induction motor running at full load is 39 A at 0.75 power factor lagging. Calculate the intake taken from the supply (a) in kW, (b) in kVA.

   Find what size capacitor connected across the motor terminals would cause the intake in kVA to be equal to the power in kW. (CGLI)

4. A group of single-phase motors takes 50 A at 0.4 power factor lagging from a 230 V supply. Calculate the apparent power and the true power input to the motors. Determine also the leading kVAr to be taken by a capacitor in order to improve the power factor to 0.8 lagging.

5. A welding set takes 60 A from a 230 V a.c. supply at 0.5 p.f. lagging. Calculate its input in (a) kVA, (b) kW.

   Determine the kVAr rating of a capacitor which will improve the power factor to 0.9 lagging. What total current will now flow?

6. Explain with the aid of a phasor diagram the meaning of power factor in the alternating-current circuit. Why is a low power factor undesirable?

A single-phase load of 20 kW at a power factor of 0.72 is supplied at 240 V a.c. Calculate the decrease in current if the power factor is changed to 0.95 with the same kW loading.   (CGLI)

7. An induction motor takes 13 A from the 240 V single-phase 50 Hz a.c. mains at 0.35 p.f. lagging. Determine the value of the capacitor in microfarads which, when connected in parallel with the motor, will improve the power factor to 0.85 lagging. Find also the supply current at the new power factor.

8. A consumer's load is 100 kVA at 0.6 p.f. lagging from a 240 V 50 Hz supply. Calculate the value of capacitance required to improve the power factor as shown in the table below:

Power factor	0.7	0.75	0.8	0.85	0.9	0.95	1.0
Capacitance required (μF)							

9. An appliance takes a current of 45 A at 0.2 power factor lagging. Determine the current to be taken by a bank of capacitors in order to improve the power factor to 0.6 lagging. Calculate the value of the capacitors in microfarads if they are supplied at (a) 240 V, (b) 415 V, and the supply frequency is 50 Hz.

10. A test on a mercury vapour lamp gave the following results:
    Without power-factor-improvement capacitor:
    volts 230   amperes 2.22   watts 260
    With power-factor-improvement capacitor:
    volts 230   amperes 1.4   watts 260
    The supply frequency was 50 Hz. Calculate the value of the capacitor in microfarads.

11. A transformer is rated at 10 kVA 230 V. The greatest current it can supply at 0.8 p.f. is
    (a) 43.3 A   (b) 34.8 A   (c) 23 A   (d) 230 A

12. The power output of the transformer of question 11 at 0.8 p.f. is
    (a) 8 kW   (b) 12.5 kW   (c) 19.2 kW   (d) 3 kW

13. A single-phase circuit supplies a load of 20 kVA at 0.8 p.f. lagging. The kVAr rating of a capacitor to improve the power factor to unity is
    (a) 16   (b) 12   (c) 25   (d) 33.3

14. In order to improve the power factor, a circuit requires a capacitor to provide 6 kVAr at 230 V 50 Hz. Its value in microfarads is
    (a) 1430 μF   (b) 143 μF   (c) 346 μF   (d) 3460 μF

Three-phase circuit calculations

Figure 66 shows three loads connected in the star formation to a three-phase four-wire supply system.

Figure 67 shows the phasor diagram. The voltage between the red line (or red phase conductor) and neutral, U_{RN}, is taken as reference. The phase sequence is red, yellow, blue, so the other line-to-neutral voltages (or *phase* voltages) lie as shown.

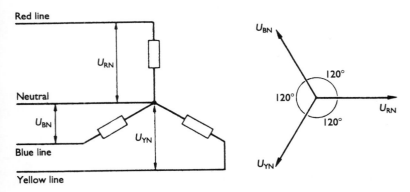

Fig. 66   **Fig. 67**

If $U_{RN} = U_{YN} = U_{BN}$ and they are equally spaced, the system of *voltages* is balanced.

Let U_L be the voltage between any pair of lines (the *line* voltage) and

$$U_P = U_{RN} = U_{YN} = U_{BN} \text{ (the } phase \text{ voltage)}$$

then $U_L = \sqrt{3}U_P$

and $I_L = I_P$

where I_L is the current in any line and I_P is the current in any load or phase.

The power per phase is $P = U_P I_P \cos \phi$ and the total power is the sum of the amounts of power in each phase.

If the currents are equal and the phase angles are the same, as in Fig. 68, the load on the system is balanced and, except in special circumstances, the current in the neutral is zero. The total power is

$$P = \sqrt{3} U_L I_L \cos \phi$$

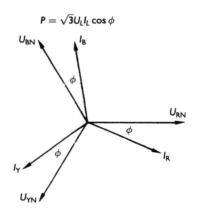

Fig. 68

EXAMPLE I   An inductor of 0.3 H and resistance $60 \, \Omega$ is connected between each line and the neutral of a 50 Hz three-phase system. The voltage between pairs of lines is 400 V. Calculate (a) the current in each line, (b) the total power in the system.

(a) Figure 66 illustrates the connection, where each block represents an inductor of impedance

$$X_L = 2\pi f L$$
$$= 2\pi \times 50 \times 0.3$$
$$= 94.24 \, \Omega$$

For each inductor, the impedance Z between each line and neutral is given by

$$Z^2 = R^2 + X_L^2$$
$$= 60^2 + 94.24^2$$
$$\therefore \quad Z = \sqrt{(60^2 + 94.24^2)}$$
$$= 112\,\Omega$$

Phase voltage $= \dfrac{400}{\sqrt{3}} = 230\,\text{V}$

Current in each line $= I_L = \dfrac{230}{112}$
$$= 2.05\,\text{A}$$

(b) The impedance triangle for the inductor may now be drawn (Fig. 69).

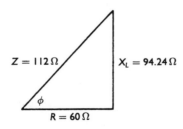

Fig. 69

The power factor of the inductor is given by

$$\cos\phi = \frac{R}{Z}$$
$$= \frac{60}{112} = 0.535 \text{ (lagging)}$$

The power in the system is

$$P = \sqrt{3}U_L I_L \cos\phi$$
$$= \sqrt{3} \times 400 \times 2.05 \times 0.535$$
$$= 760\,\text{W}$$

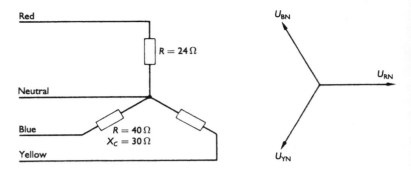

Fig. 70

EXAMPLE 2   The following loads are connected to a $400\,\text{V}$ three-phase four-wire system (Fig. 70):

between the red line and neutral, a non-inductive resistor of $24\,\Omega$;

between the yellow line and neutral, $886\,\text{W}$ at 0.555 p.f. lagging;

between the blue line and neutral, a capacitor of reactance $30\,\Omega$ in series with a resistor of $40\,\Omega$.

The phase sequence is red, yellow, blue. Calculate (a) the current in each line, (b) the total power.

(a)   Line current = phase current

$$= \frac{\text{line-to-neutral voltage}}{\text{impedance between line and neutral}}$$

Line-to-neutral voltage $= U_\text{P} = \dfrac{1}{\sqrt{3}} \times U_\text{L}$

$$= \frac{400}{\sqrt{3}}$$

$$= 230\,\text{V}$$

Current in red line $I_\text{R} = \dfrac{230}{24}$

$$= 9.58\,\text{A}$$

The power in the yellow-phase circuit is

$$P_\text{Y} = U_\text{YN} I_\text{Y} \cos \phi_\text{Y}$$

$\therefore$ $886 = 230 \times I_\text{Y} \times 0.555$

$$\therefore \quad \text{current in yellow line } I_Y = \frac{886}{230 \times 0.555}$$

$$= 6.94 \, \text{A}$$

$$\text{Current in blue line } I_B = \frac{230}{\sqrt{(R^2 + X_C{}^2)}}$$

$$= \frac{230}{\sqrt{(40^2 + 30^2)}}$$

$$= 4.6 \, \text{A}$$

(b) The phase angle between this current and the blue-to-neutral voltage is given by

$$\tan \phi_B = \frac{X_C}{R} = \frac{30}{40} = 0.75$$

$$\therefore \qquad \phi_B = 36° \, 52' \quad (\text{lead})$$

$$\text{Power in the red phase } P_R = 230 \times 10$$

$$= 2300 \, \text{W}$$

$$\text{Power in the blue phase } P_B = 230 \times 4.6 \times \cos 36° \, 52'$$

$$= 230 \times 4.6 \times 0.8$$

$$= 846.4 \, \text{W}$$

$$\text{Total power } P = 2300 + 886 + 846.4$$

$$= 4032 \, \text{W}$$

DELTA OR MESH CONNECTION

Figure 71 shows three loads connected in the delta or mesh formation to a three-phase supply system.

Figure 72 shows the phasor diagram of the line voltages, with the red-to-yellow voltage taken as reference.

The voltage applied across any load is the line voltage U_L, and the line current is the vector difference between the currents in the two loads connected to that line. In particular, if the load currents are all equal and make equal phase angles with their respective voltages, the system is balanced and

$$I_L = \sqrt{3} I_P$$

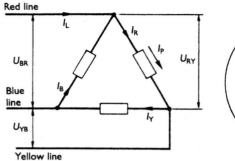

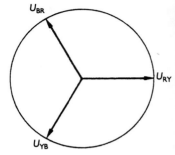

Fig. 71   **Fig. 72**

The total power under these conditions is

$$P = \sqrt{3}U_{\mathrm{L}}I_{\mathrm{L}}\cos\phi$$

EXAMPLE Three coils, each of resistance $40\,\Omega$ and inductive reactance $30\,\Omega$, are connected in delta to a 400 V three-phase system. Calculate (a) the current in each coil, (b) the line current, (c) the total power.

The circuit diagram is as Fig. 71.

(a) Impedance of each coil, $Z = \sqrt{(R^2 + X_L{}^2)}$

$$= \sqrt{(40^2 + 30^2)}$$

$$= 50\,\Omega$$

$$\text{Current in each coil} = \frac{U}{Z}$$

$$= \frac{400}{50}$$

$$= 8\,\mathrm{A}$$

(b) Line current $I_{\mathrm{L}} = \sqrt{3} \times 8$

$$= 13.86\,\mathrm{A}$$

(c) The power factor of the coil is

$$\cos \phi = \frac{R}{Z}$$

$$= \frac{40}{50} = 0.8$$

Total power $P = \sqrt{3}U_L I_L \cos \phi$

$$= \sqrt{3} \times 400 \times 13.86 \times 0.8$$

$$= 7682\,\text{W}$$

EXERCISE 11

1. Three equal coils of inductive reactance $30\,\Omega$ and resistance $40\,\Omega$ are connected in star to a three-phase supply of line voltage $400\,\text{V}$. Calculate the line current and the total power.

2. The load connected between each line and the neutral of a $400\,\text{V}$ $50\,\text{Hz}$ three-phase circuit consists of a capacitor of $31.8\,\mu\text{F}$ in series with a resistor of $100\,\Omega$. Calculate the current in each line and the total power.

3. The load connected between each line and the neutral of a $400\,\text{V}$ three-phase supply consists of:

   between the red line and neutral, non-inductive resistance of $25\,\Omega$;
   between the yellow line and neutral, inductive reactance $12\,\Omega$ in series with resistance $5\,\Omega$;
   between the blue line and neutral, capacitive reactance $17.3\,\Omega$ in series with resistance $10\,\Omega$.

   Calculate the current in each line and the total power.

4. A star-connected resistance bank, each resistor of $30\,\Omega$, is connected to a $400\,\text{V}$ three-phase supply. Connected in parallel to the same supply is a star-connected capacitor bank, each capacitor having reactance $40\,\Omega$. Calculate the resultant current in each line and the total power.

5. Three capacitors, each of $10\,\Omega$ reactance, are to be connected to $400\,\text{V}$ three-phase mains for the power-factor improvement. Calculate the currents in each line taken if they are connected (a) in star, (b) in mesh.

6. A 440 V, three-phase, four-wire system supplies a balanced load of 10 kW. Three single-phase resistive loads are now added between lines and neutral as follows: R–N 2 kW, Y–N 4 kW, B–N 3 kW. Find the current in each line.

7. Three 30 Ω resistors are connected (a) in star, (b) in delta to a 400 V three-phase system. Calculate the current in each resistor, the line currents, and the total power for each connection.

8. Each branch of a mesh-connected load consists of resistance 20 Ω in series with inductive reactance 30 Ω. The line voltage is 400 V. Calculate the line currents and the total power.

9. Three coils, each with resistance 45 Ω and with inductance 0.2 H, are connected to a 400 V three-phase supply at 50 Hz, (a) in mesh, (b) in star. For each method of connection calculate (i) the current in each coil, (ii) the total power in the circuit.

10. A three-phase load consists of three similar inductive coils, each of resistance 50 Ω and inductance 0.3 H. The supply is 400 V 50 Hz. Calculate

    the current in each line,

    the power factor,

    the total power

when the load is (a) star-connected, (b) delta-connected.

11. Three equal resistors are required to absorb a total of 24 kW from a 400 V three-phase system. Calculate the value of each resistor when they are connected (a) in star, (b) in mesh.

12. To improve the power factor, a certain installation requires a total of 48 kVAr equally distributed over the three phases of a 415 V 50 Hz system. Calculate the value of the capacitors required (in microfarads) when the capacitors are connected (a) in star, (b) in delta.

13. The following loads are connected to a three-phase three-wire 400 V 50 Hz supply system:

    between red and yellow lines, non-inductive resistance 60 Ω;

    between yellow and blue lines, a coil of inductive reactance 30 Ω and resistance 40 Ω;

    between blue and red lines, a capacitor of 100 μF.

Calculate the current through each load and the total power.

14. A motor–generator set consists of a d.c. generator driven by a three-phase a.c. motor. The generator is 65% efficient and delivers 18 A at 220 V.

The motor is 75% efficient and operates at 0.5 p.f. lagging from a 415 V supply. Calculate (a) the power output of the driving motor, (b) the line current taken by the motor.

15. A conveyor raises 1600 kg of goods through a vertical distance of 5 m in 20 s and it is driven through a gear which is 55% efficient. Calculate the power output of the motor required for this work.

   If a three-phase 400 V motor having an efficiency of 78% is fitted, calculate the line current. Assume a power factor of 0.7.

16. A 415 V three-phase star-connected alternator supplies a delta-connected induction motor of full-load efficiency 87% and power factor 0.8 which delivers 14 920 W. Calculate:
   (a) the current in each motor winding;
   (b) the current in each alternator winding;
   (c) the power to be developed by the engine driving the alternator, assuming that the efficiency of the alternator is 82%.

17. A three-phase transformer supplies a block of flats at 250 V line-to-neutral. The load is balanced and totals 285 kW at 0.95 power factor. The turns ratio of the transformer, primary-to-secondary, is 44 : 1 and the primary is connected in mesh.
   (a) What is the primary line voltage?
   (b) Draw a connection diagram and mark in the values of phase and line currents in the primary and secondary.

18. On the largest possible axis, plot sine waves to represent three alternating voltages each having a maximum value of 100 V and each displaced from the next by 120°. Add corresponding ordinates to show that the instantaneous resultant value of the three waves is always zero. (This confirms that with simple sinusoidal voltages or currents in a balanced three-phase system the neutral carries no current.)

19. Using the same size axes as in question 18, construct the third harmonic of each of the voltages (this is, in each case, simply a sine wave of frequency three times that used in question 18). Add corresponding ordinates to show that the instantaneous resultant is not zero. (This shows that in three-phase circuits supplying loads which generate third-harmonic currents, e.g. fluorescent-lighting systems, the neutral may carry a significant current.)

Voltmeters and ammeters: changing the use and extending the range

The voltmeter is a high-resistance instrument, and its essential electrical features may be represented by the equivalent circuit of Fig. 73.

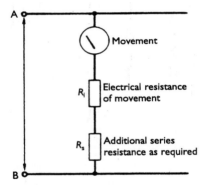

Fig. 73

R_i is the 'internal resistance' of the movement, i.e. the resistance of the moving coil or the resistance of the fixed coil in the case of a moving-iron instrument.

Independently of its resistance, the movement will require a certain current to deflect the pointer across the full extent of the scale against the effect of the controlling springs. This is the current required for full-scale deflection (f.s.d.).

The range of voltage which the instrument can indicate is governed by the total resistance R as measured between the terminals A and B, and

$$R = R_i + R_s$$

If $I_{f.s.d.}$ is the current required to produce full-scale deflection and R is the *total* resistance between A and B, the voltage between A and B at full-scale deflection is

$$U = R \times I_{f.s.d.}$$

$I_{f.s.d.}$ is fixed by the mechanical and electrical characteristics of the instrument and is not normally variable. The resistance R, however, can be fixed at any convenient value by adding the additional series resistance (R_s) as required.

EXAMPLE   An instrument has internal resistance $20\,\Omega$ and gives f.s.d. with a current of 1 mA. Calculate the additional series resistance required to give f.s.d. at a voltage of 100 V.

$$U = R \times I_{f.s.d.}$$

$\therefore$ $100 = R \times \dfrac{1}{1000}$ (note the conversion of milliamperes to amperes)

$\therefore$ $R = 100 \times 1000\,\Omega$

$ = 100\,000\,\Omega$

But the instrument has internal resistance of $20\,\Omega$; thus the additional resistance required is

$$R_s = R - R_i$$
$$= 100\,000 - 20$$
$$= 99\,980\,\Omega$$

(Note that this is a somewhat unrealistic value in terms of what it is economically practical to manufacture. In practice, the additional resistance would be constructed to the nominal value of $100\,000\,\Omega$ ($100\,k\Omega$) and slight adjustments would be made as necessary at the calibration stage to obtain f.s.d. with an applied 100 V.)

Any applied voltage less than 100 V of course produces a corresponding lower reading on the instrument.

The additional series resistor R_s is also known as a *multiplier*.

The ammeter is a low-resistance instrument, and its equivalent electrical circuit is shown in Fig. 74.

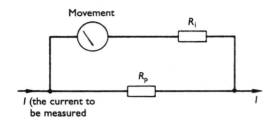

Movement

R_i

R_p

I (the current to be measured

I

Fig. 74

The resistor R_p connected in parallel is the 'shunt' through which most of the current to be measured flows. Its value will be low compared with the internal resistance of the movement R_i. The calculation of R_p proceeds as follows.

Knowing R_i and the current required to give f.s.d., determine the voltage required to produce f.s.d. of the movement. For example, using the information of the previous example ($I_{f.s.d.} = 1\,\text{mA}$ and $R_i = 20\,\Omega$),

$$\text{p.d. required for f.s.d.} = U_{f.s.d.} = I_{f.s.d.} \times R_i$$

$$= \frac{1}{1000} \times 20$$

$$= \frac{20}{1000}\,\text{V}$$

and this is the voltage drop which must be produced across the shunt resistor R_p.

The current which flows through R_p is the current to be measured minus the current which flows through the movement. If the greatest value of current to be measured is 20 A, the current which flows through R_p is $I = 20\,\text{A} - 1\,\text{mA} = 19.999\,\text{A}$. The voltage across R_p is then $V_{f.s.d.} = (20/1000)\,\text{V}$,

$$\therefore \quad R_\text{p} = \frac{U_\text{f.s.d.}}{I}$$

$$= \frac{20/1000}{19.999}$$

$$= 0.001\,\Omega$$

In fact, no significant difference is made if the 1 mA of total current which flows through the movement and not through the shunt is ignored in the calculation of R_p. Again, it is usual to manufacture the shunt to the nominal value calculated above and then to make slight adjustments at the calibration stage to obtain the desired full-scale deflection.

EXAMPLE   A moving-coil instrument gives full-scale deflection with a current of 1.2 mA, and its coil has resistance 40 Ω. Determine
(a) the value of the multiplier required to produce a voltmeter reading up to 50 V,
(b) the value of the shunt required to convert the instrument to an ammeter reading up to 10 A.

(a) Total resistance required to restrict the current to 1.2 mA from a 50 V supply is

$$R = \frac{50\,\text{V}}{(1.2/1000)\,\text{A}} \quad \text{(note the conversion to amperes)}$$

$$= 41\,667\,\Omega$$

The accurate value of additional series resistance required is

$$R_\text{s} = 41\,667 - 40$$

$$= 41\,627\,\Omega$$

(b) $\begin{aligned}\text{Voltage required}\\ \text{to produce f.s.d.}\end{aligned} = \frac{1.2}{1000} \times 40$

$$= \frac{4.8}{1000}\,\text{V}$$

Then $\qquad \dfrac{4.8\,\text{V}}{1000} = R_\text{p} \times (10\,\text{A} - 0.0012\,\text{A})$

$$\therefore \qquad R_\text{p} = \frac{4.8}{1000 \times 9.9988}$$

$$= 4.8 \times 10^{-4}\,\Omega$$

Again, the 1.2 mA of current which flows through the instrument could have been neglected in calculating R_p.

EXERCISE 12

1. The coil of a moving-coil instrument has resistance 50 Ω, and a current of 0.8 mA is required to produce full-scale deflection. Calculate the voltage required to produce full-scale deflection.

2. A moving-coil instrument movement was tested without either shunt or multiplier fitted and it was found that at full-scale deflection the current through the coil was 1.15 mA and the voltage across it was 52 mV. Determine the resistance of the coil.

3. A moving-coil instrument gives full-scale deflection with a current of 1.5 mA and has resistance (without shunt or multiplier) of 25 Ω. Determine the value of additional series resistance (the multiplier) required to produce a voltmeter capable of measuring up to 150 V.

4. Using the instrument movement of question 3, modify the meter to measure currents up to 25 A by calculating the value of a suitable shunt resistor.

5. Given an instrument movement of resistance 40 Ω and requiring a current of 1 mA to produce f.s.d., determine the values of the various resistors required to produce a multi-range instrument having the following ranges:
   Voltage: 0–10 V, 0–150 V, 0–250 V
   Current: 0–1 A, 0–10 A

6. A moving-coil instrument requires 0.75 mA of current to produce f.s.d. at a voltage of 50 mV. The resistance of the coil is
   (a) 0.015 Ω (b) 0.067 Ω (c) 66.7 Ω (d) 66 700 Ω

7. The coil of a moving-coil instrument has resistance 45 Ω and requires a current of 1.15 mA to produce f.s.d. The p.d. required to produce f.s.d. is
   (a) 51.8 V   (b) 39 V   (c) 0.025 V   (d) 51.8 V

8. The value of the multiplier required to convert the instrument of question 7 to a voltmeter to measure up to 250 V is approximately
   (a) 217 kΩ (b) 6.25 Ω (c) 217 Ω (d) 288 kΩ

9. The approximate value of the shunt required to convert the instrument of question 6 to an ammeter to measure up to 25 A is
   (a) 2 Ω (b) 0.002 Ω (c) 0.067 Ω (d) 0.125 Ω

Lighting calculations

When a lamp of luminous intensity I candela in all directions
below the horizontal is suspended d metres above a surface
(Fig. 75), the illuminance produced at a point P on the surface
below the lamp is given by

$$\text{illuminance } E_P = \frac{I}{d^2} \text{ lumen per m}^2 = \frac{I}{d^2} \text{ lux (lx)}$$

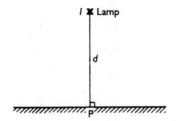

Fig. 75

EXAMPLE 1   A luminaire producing luminous intensity 1500
candela in all directions below the horizontal is suspended 4 m
above the floor. Calculate the illuminance produced at a point P
immediately below the luminaire.

The illuminance at P is

$$E_P = \frac{I}{d^2}$$
$$= \frac{1500}{4^2}$$
$$= 93.8 \, \text{lx}$$

EXAMPLE 2   If the luminaire of Example 1 were to be raised by
1 m, what would be the new illuminance at P?

The new illuminance at P is

$$E_P = \frac{I}{d^2}$$

$$= \frac{1500}{(4+1)^2} \quad \text{(new value of } d\text{)}$$

$$= \frac{1500}{5^2}$$

$$= 60 \, \text{lx}$$

POINT-BY-POINT LIGHTING CALCULATIONS USING THE COSINE LAW

When a lamp of luminous intensity I candela in all directions below the horizontal is suspended above a horizontal surface, the illuminance produced at any point Q on the surface (Fig. 76) is given by

$$\text{illuminance } E_Q = \frac{I}{h^2} \cos \theta \text{ lux}$$

where h and θ are as shown in Fig. 76.

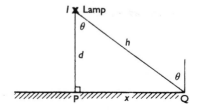

Fig. 76

Also $\quad h^2 = d^2 + x^2 \quad$ (Pythagoras)

and $\quad \cos \theta = \dfrac{d}{h}$

$$= \frac{d}{\sqrt{(d^2 + x^2)}}$$

EXAMPLE I   For the same conditions as in Example 1 in the previous section, calculate the illuminance at a point Q 2.5 m away from P in a horizontal line.

The illuminance at Q is

$$E_Q = \frac{I}{h^2} \cos \theta$$

As shown above,

$$
\begin{aligned}
h^2 &= d^2 + x^2 \\
&= 4^2 + 2.5^2 \\
&= 16 + 6.25 \\
&= 22.25
\end{aligned}
$$

$$
\begin{aligned}
\cos \theta &= \frac{d}{h} \\
&= \frac{4}{\sqrt{22.25}} \\
&= 0.8433
\end{aligned}
$$

$$
\begin{aligned}
\therefore \quad E_Q &= \frac{1500}{22.25} \times 0.8433 \\
&= 56.8\,\text{lx}
\end{aligned}
$$

EXAMPLE 2 Two lamps are suspended 10 m apart and at a height of 3.5 m above a surface (Fig. 77). Each lamp emits 350 cd. Calculate
(a) the illuminance on the surface midway between the lamps,
(b) the illuminance on the surface immediately below each of the lamps.

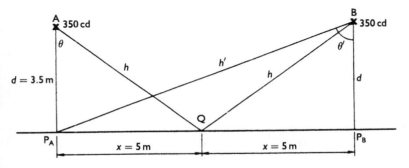

Fig. 77

(a) For one lamp, the illuminance at Q is

$$E_Q = \frac{I}{h^2} \cos\theta$$

$$= \frac{350}{3.5^2 + 5^2} \times \frac{3.5}{\sqrt{(3.5^2 + 5^2)}}$$

$$= \frac{350}{12.25 + 25} \times \frac{3.5}{\sqrt{(12.25 + 25)}}$$

$$= \frac{350}{37.25} \times \frac{3.5}{\sqrt{37.25}}$$

$$= 5.388\,\text{lx}$$

The illuminance from two lamps is double that due to one lamp, since the conditions for both lamps are identical. Thus

$$\text{total illuminance at Q} = 2 \times 5.388$$

$$= 11.8\,\text{lx}$$

(b) At P_A below lamp A, the illuminance due to lamp A is

$$E_{P_A} = \frac{I}{d^2}$$

$$= \frac{350}{3.5^2}$$

$$= 28.57\,\text{lx}$$

In calculating the illuminance at P_A due to lamp B, we have a new distance h', a new distance x', and a new angle θ' to consider.

$$x' = 2x$$

$$= 10$$

$$(h')^2 = 3.5^2 + 10^2$$

$$= 112.25$$

$$\therefore \qquad h' = 10.59$$

$$\cos\theta' = \frac{d}{h'}$$

$$= \frac{3.5}{10.59}$$

$$= 0.331$$

∴ illuminance at P_A due to lamp B is

$$E_{P_B} = \frac{350}{112.25} \times 0.331$$
$$= 1.013$$

Total illuminance at $P_A = 28.57 + 1.013$
$$= 29.6\,\text{lx}$$

and, as the conditions at P_B are the same as those at P_A, this will also be the illuminance below lamp B.

EXERCISE 13

1. A lamp emitting 250 cd in all directions below the horizontal is fixed 4 m above a horizontal surface. Calculate the illuminance at (a) a point P on the surface vertically beneath the lamp, (b) a point Q 3 m away from P.

2. Two luminaires illuminate a passageway. The luminaires are 12 m apart. Each emits 240 cd and is 3 m above the floor. Calculate the illuminance at a point on the floor midway between the luminaires.

3. Determine the illuminance at a point vertically beneath one of the luminaires in question 2.

4. An incandescent filament luminaire is suspended 2 m above a level work bench. The luminous intensity in all directions below the horizontal is 400 candelas.

   Calculate the illuminance at a point A on the surface of the bench immediately below the luminaire, and at other bench positions 1 m, 2 m, and 3 m from A in a straight line. Show the values on a suitable diagram.   (CGLI)

5. Two incandescent filament luminaires are suspended 2 m apart and 2.5 m above a level work bench. The luminaires give a luminous intensity of 200 candelas in all directions below the horizontal. Calculate the total illuminance at bench level, immediately below each luminaire and midway between them.

6. A work bench is illuminated by a luminaire emitting 350 cd in all directions below the horizontal and mounted 2.5 m above the working surface.

   (a) Calculate the illuminance immediately below the luminaire.

   (b) It is desired to increase the illuminance by 10%. Determine two methods of achieving this, giving calculated values in each case.

7. A lamp emitting 450 cd in all directions is suspended 3 m above the floor. The illuminance on the floor immediately below the lamp is

(a) 150 lx   (b) 1350 lx   (c) 50 lx   (d) 0.02 lx

8. If the lamp of question 7 is reduced in height by 0.5 m, the illuminance produced immediately below it is

(a) 72 lx   (b) 36.7 lx   (c) 129 lx   (d) 180 lx

Electromagnetic induction

INDUCED E.M.F.

When a conductor moves through a magnetic field in a direction at right angles to the field, the e.m.f. induced is

$$e = Blv \text{ volts}$$

where B is the flux density (T),

l is the effective length of conductor (m),

and v is the velocity of the conductor (m/s).

EXAMPLE I   The e.m.f. induced in a conductor of effective length 0.25 m moving at right angles through a magnetic field at a velocity of 5 m/s is 1.375 V. Calculate the magnetic flux density.

$$e = Blv$$

$$\therefore \quad 1.375 = B \times 0.25 \times 5$$

$$\therefore \quad B = \frac{1.375}{0.25 \times 5}$$

$$= 1.1 \text{ T}$$

If the flux linking a coil of N turns is Φ_1 (Wb) at some instant of time t_1 and Φ_2 (Wb) at another instant of time t_2, the e.m.f.

induced is

$$e = \text{number of turns} \times \text{rate of change of flux}$$

$$= N \times \frac{\text{change in flux}}{\text{time required for change in flux}}$$

$$= N \times \frac{\Phi_2 - \Phi_1}{t_2 - t_1} \text{ volts}$$

EXAMPLE 2   The flux linking a coil of 50 turns changes from 0.042 Wb to 0.075 Wb in 0.003 seconds. Calculate the e.m.f. induced.

$$e = N \times \frac{\Phi_2 - \Phi_1}{t_2 - t_1}$$

Here $t_2 - t_1 = 0.003$

thus $e = 50 \times \dfrac{0.075 - 0.042}{0.003}$

$$= 550 \text{ V}$$

SELF-INDUCTANCE

If the self-inductance of a magnetic system is L henrys and the current changes from I_1 at time t_1 to I_2 at time t_2, the induced e.m.f. is

$$e = L \times \text{rate of change of current}$$

$$= L \times \frac{I_2 - I_1}{t_2 - t_1} \text{ volts}$$

where the current is in amperes and the time in seconds.

EXAMPLE I   A coil has self-inductance 3 H, and the current through it changes from 0.5 A to 0.1 A in 0.01 s. Calculate the e.m.f. induced.

$$e = L \times \text{rate of change of current}$$

$$= 3 \times \frac{0.5 - 0.1}{0.01}$$

$$= 120 \text{ V}$$

The self-inductance of a magnetic circuit is given by

$$\text{self-inductance} = \frac{\text{change in flux linkage}}{\text{corresponding change in current}}$$

$$L = N \times \frac{\Phi_2 - \Phi_1}{I_2 - I_1} \text{ henrys}$$

where N is the number of turns on the magnetizing coil and Φ_2, I_2; Φ_1, I_1 are corresponding values of flux and current.

EXAMPLE 2 The four field coils of a d.c. machine each have 1250 turns and are connected in series. The change in flux produced by a change in current of 0.25 A is 0.0035 Wb. Calculate the self-inductance of the system.

$$L = N \times \frac{\Phi_2 - \Phi_1}{I_2 - I_1}$$

$$= 4 \times 1250 \times \frac{0.0035}{0.25}$$

$$= 70 \text{ H}$$

MUTUAL INDUCTANCE

If two coils A and B have mutual inductance M henrys, the e.m.f. in coil A due to current change in coil B is

$$e_A = M \times \text{rate of change of current in coil B}$$

Thus, if the current in coil B has values I_1 and I_2 at instants of time t_1 and t_2,

$$e = M \times \frac{I_2 - I_1}{t_2 - t_1} \text{ volts}$$

EXAMPLE 1 Two coils have mutual inductance 3 H. If the current through one coil changes from 0.1 A to 0.4 A in 0.15 s, calculate the e.m.f. induced in the other coil.

$$e = 3 \times \frac{0.4 \times 0.1}{0.15} \quad (t_2 - t_1 = 0.15)$$

$$= 6 \text{ V}$$

The mutual inductance between two coils is given by

$$M = N_A \times \frac{\Phi_2 - \Phi_1}{I_{B1} - I_{B2}} \text{ henrys}$$

where N_A is the number of turns on coil A and Φ_2 and Φ_1 are the values of flux linking coil A due to the two values of current in coil B, I_{B2} and I_{B1} respectively.

EXAMPLE 2 The secondary winding of a transformer has 200 turns. When the primary current is 1 A the total flux is 0.05 Wb, and when it is 2 A the total flux is 0.095 Wb. Assuming that all the flux links both windings, calculate the mutual inductance between the primary and secondary.

$$M = N_A \times \frac{\Phi_2 - \Phi_1}{I_{B1} - I_{B2}}$$

$$= 200 \times \frac{0.095 - 0.05}{2 - 1}$$

$$= 9 \text{ H}$$

EXERCISE 14

1. A conductor of effective length 0.2 m moves through a uniform magnetic field of density 0.8 T with a velocity of 0.5 m/s. Calculate the e.m.f. induced in the conductor.

2. Calculate the velocity with which a conductor 0.3 m long must pass at right angles through a magnetic field of flux density 0.65 T in order that the induced e.m.f. shall be 0.5 V.

3. Calculate the e.m.f. induced in a coil of 1200 turns when the flux linking with it changes from 0.03 Wb to 0.045 Wb in 0.1 s.

4. The magnetic flux in a coil of 850 turns is 0.015 Wb. Calculate the e.m.f. induced when this flux is reversed in 0.25 s.

5. A coil has self-inductance 0.65 H. Calculate the e.m.f. induced in the coil when the current through it changes at the rate of 10 A/s.

6. A current of 5 A through a certain coil is reversed in 0.1 s, and the induced e.m.f. is 15 V. Calculate the self-inductance of the coil.

7. A coil has 2000 turns. When the current through the coil is 0.5 A the flux is 0.03 Wb; when the current is 0.8 A the flux is 0.045 Wb. Calculate the self-inductance of the coil.

8. An air-cored coil has 250 turns. The flux produced by a current of 5 A is 0.035 Wb. Calculate the self-inductance of the coil. (Hint: in an air-cored coil, current and magnetic flux are directly proportional. When there is no current, there is no flux.)

9. Two coils have mutual inductance 2 H. Calculate the e.m.f. induced in one coil when the current through the other changes at the rate of 25 A/s.

10. Two coils have mutual inductance 0.15 H. At what rate must the current through one change in order to induce an e.m.f. of 10 V in the other?

11. Two coils are arranged so that the same flux links both. One coil has 1200 turns. When the current through the other coil is 1.5 A, the flux is 0.045 Wb; when the current is 2.5 A the flux is 0.07 Wb. Calculate the mutual inductance between the coils.

12. Calculate the e.m.f. induced in one of the coils in question 11 if a current of 0.2 A in the other coil is reversed in 0.15 s.

13. The e.m.f. induced in a conductor of length 0.15 m moving at right angles to a magnetic field with a velocity of 7.5 m/s is 22.5 mV. The magnetic flux density is

    (a) 20 T   **(b)** 25.3 T   **(c)** 0.02 T   **(d)** 0.0253 T

14. The magnetic flux linking a coil of 150 turns changes from 0.05 Wb to 0.075 Wb in 5 ms. The e.m.f. induced is

    (a) 750 V   **(b)** 0.75 V   **(c)** 37.5 V   **(d)** 37 500 V

15. When the current through a coil changes from 0.15 A to 0.7 A in 0.015 s, the e.m.f. induced is 100 V. The self-inductance of the coil is

    (a) 367 H   **(b)** 0.367 H   **(c)** 2.73 H   **(d)** 1.76 H

16. Two coils have mutual inductance 0.12 H. The current through one coil changes at the rate of 150 A/s. The e.m.f. induced in the other is

    (a) 1250 V   **(b)** 0.0008 V   **(c)** 180 V   **(d)** 18 V

Practical electrical installation calculations

APPLICATION OF DIVERSITY FACTORS

Not all the installed load in a premises, i.e. lighting, heating, cooking and power circuits, will be in full use continuously, so to minimize the size of cables and other equipment and thus achieve economy in the cost of electrical installation of premises, we may apply diversity factors to certain circuits to arrive at an assumed maximum demand for a particular electrical installation.

The *IEE On-site Guide* contains tables in Appendix 1 giving information on the assumed current demand for points of utilization and current-using equipment as well as giving allowances for final circuit diversity.

EXAMPLE 1   A domestic premises supplied at 230 V 50 Hz has the following connected load:

Cooker, assume 10 kW (control unit with socket outlet)
Hob, assume 4 kW
Direct heating, assume 6 kW
Instantaneous shower unit 8 kW
Two 30 A ring-type utility circuits
Lighting, assume 1800 W.

Determine the assumed maximum demand for the premises.

Assumed maximum demand (A.M.D.) using *IEE On-site Guide* Table 1B:

Cooker/hob $\dfrac{14\,000}{230} = 60.87\,\text{A}$

A.M.D. is 10 A plus 30% of 50.87 A
i.e. 10 A + 15.26 A = 25.26 A

Socket outlet at cooker unit

$\qquad$ A.M.D. is 5 A

Direct heating $\quad \dfrac{6\,000}{230} = 26.09\,\text{A}$

$\qquad$ A.M.D. is 10 A plus 50% of 16.09

$\qquad$ i.e. 10 A + 8 A = 18 A

Shower unit $\quad \dfrac{8000}{230} = 34.8\,\text{A}$

$\qquad$ A.M.D. is 34.8 A (no diversity)

Socket outlet circuits 2 at 30 A

$\qquad$ A.M.D. is 30 A plus 40% of 30 A

$\qquad$ i.e. 30 A + 12 A = 42 A

Lighting $\quad \dfrac{1800}{230} = 7.83\,\text{A}$

$\qquad$ A.M.D. is 60% of 7.83 A

$\qquad$ i.e. 4.7 A

Thus assumed maximum demand is

$$25.26 + 5 + 18 + 34.8 + 42 + 4.7 = 111.76\,\text{A}$$

Note As the assumed maximum demand is over 100 A, probably above the rating of the switch-disconnector of a 'domestic' consumer unit, thought could be given to providing a second consumer unit, possibly of the 'split load' type for the socket outlets, etc.; this would also allow future extension to the installation.

EXAMPLE 2 A retail jewellers has the following connected load supplied at 230 V 50 Hz:

Direct heaters $\quad 2 \times 2\,\text{kW}, 3 \times 1.5\,\text{kW}, 1 \times 1.0\,\text{kW}$
Cooker $\quad 5\,\text{kW}$ (cooker unit has socket outlet)
Water heating (thermostatic) $\quad 3\,\text{kW}$
Socket outlets $\quad 30\,\text{A ring circuit}$
Shop and window lighting $\quad 2.5\,\text{kW total.}$

Determine the assumed maximum demand.

Assumed maximum demand (A.M.D.) using *IEE On-site Guide* Table 1B (small shop premises):

Heaters

$$2 \times 2\,\text{kW} + 3 \times 1.5 + 1.0 = \frac{2000}{230} + \frac{7500}{230} \times \frac{75}{100}$$

$$= 8.7\,\text{A} + 24.46\,\text{A} = 33.16\,\text{A}$$

Cooker $\dfrac{5000}{230} = 21.7\,\text{A}$

Socket at cooker control   5 A

Water heater $\dfrac{3000}{230} = 13.04\,\text{A}$ (no diversity)

Socket outlets   30 A

Lighting $\dfrac{2500}{230} = 10.9 \times \dfrac{90}{100} = 9.8\,\text{A}$

Assumed maximum demand $= 33.16 + 21.7 + 5 + 13.04$

$$+ 30 + 9.8$$

$$= 112.7\,\text{A}$$

In this case there may be separate main control for associated circuits. Heating and shop window lighting may be on time switch/contactor controls with individual distribution boards and switch-disconnectors.

EXAMPLE 3   A small hotel supplied at 400/230 V 50 Hz has the following connected load:

100 lighting outlets
50 × 13 A socket outlets on 6 × 30 A ring circuits
30 × 1 kW convection heaters on coin-operated meters
16 kW thermal storage central heating boiler
Cooking equipment – 2 × 14 kW cookers, 1 × 4 kW hot cupboard, 3 × 2 kW fryers, 4 × 600 W microwave ovens plus 5 kW machines.

Determine the assumed maximum demand.

Using Tables 1A and 1B in the *IEE On-site Guide*:

100 lighting points = 100 × 100 W = 10 kW

so $\quad$ A.M.D. $= \dfrac{10\,000}{230} \times \dfrac{75}{100} = 32.6\,\text{A}$

$6 \times 30\,\text{A}$ ring circuits:

$$\text{A.M.D. is } 30\,\text{A} + \dfrac{150 \times 50}{100} = 105\,\text{A}$$

$30 \times 1\,\text{kW}$ convection heater:

$$\text{A.M.D. is } \dfrac{1000}{230} + \dfrac{29\,000 \times 50}{230 \times 100} = 4.35\,\text{A} + 63\,\text{A} = 67.35\,\text{A}$$

$16\,\text{kW}$ thermal storage:

$$\text{A.M.D. is } \dfrac{16\,000}{230} = 69.56\,\text{A} \quad \text{(no diversity)}$$

Cooking equipment

$$\dfrac{14\,000}{230} + \dfrac{14\,000 \times 80}{230 \times 100} + \dfrac{4000 + 6000 + 2400 + 5000 \times 60}{230 \times 100}$$
$$= 60.9\,\text{A} + 48.7\,\text{A} + 45.4\,\text{A} = 155\,\text{A}$$

Assumed maximum demand

$$= 32.6\,\text{A} + 105\,\text{A} + 67.35\,\text{A} + 69.56\,\text{A} + 155\,\text{A}$$
$$= 429.51\,\text{A}$$

Assuming that load is balanced over the three-phase supply then load would be approximately $143\,\text{A}$ per phase.

EXERCISE 15

1. A bungalow is supplied at 230 V 50 Hz and has the following connected load:

   18 ceiling mounted lighting outlets

   $12 \times 2\,\text{A}$ socket outlets for local luminaires

   $3 \times 30\,\text{A}$ socket outlet ring circuits

   $1 \times 10\,\text{kW}$ cooker (control unit without socket outlet)

   $1 \times 5.5\,\text{kW}$ hob unit

   $10\,\text{kW}$ of thermal storage space heating

   $1 \times 3\,\text{kW}$ immersion heater (thermostat controlled)

   $1 \times 8\,\text{kW}$ shower unit.

   Determine the assumed maximum demand, and comment upon the magnitude of this.

2. A ladies hairdressing salon is supplied at 230 V 50 Hz and has the following connected load:

    4 kW thermal storage space heating

    6 × 3 kW under-sink instantaneous water heater

    2 × 30 A socket outlet ring circuits

    2 kW of shop lighting

    2 × 500 W tungsten-halogen shop front luminaires.

Determine the assumed maximum demand and comment upon its magnitude.

3. A country hotel is at present supplied at 230 V 50 Hz and is to be rewired employing the following installation circuits:

    Lighting: four floors each having 1000 W loading

    Heating: three upper floors each having 6 × 1 kW convection heaters

         ground floor 3 × 3 kW and 3 × 2 kW convection heaters

    Socket outlets: 4 × 30 A ring circuits

    Cooking appliances: 1 × 10 kW cooker, 1 × 6 kW hob unit, 4 kW of assorted appliances (cooker control without socket outlet)

    Outside lighting: 3 × 500 W tungsten halogen floodlights.

Determine the assumed maximum demand and comment upon its magnitude.

4. An insurance office is supplied at 400/230 V 50 Hz and has the following connected load:

    4 × 30 A socket outlet circuits for computer use

    1 × 30 A socket outlet circuit for general use

    1.5 kW of fluorescent lighting

    1.0 kW of tungsten lighting

    1 × 6 kW cooker

    2 × 600 W microwave cookers

    2 × 3 kW instantaneous type hand washers

    2 × 2 kW hand dryers.

Determine the assumed maximum demand.

CURRENT RATING OF CABLES AND VOLTAGE DROP

The selection of the type of cable for an electrical installation circuit depends upon many factors, predominantly concerned with the environmental conditions; the actual selection of a cable type for a particular circuit is outside the sphere of this book and indeed for a specialized installation perhaps beyond the industrial experience and responsibility of the reader at this stage of their working life. However, site personnel are often called upon to determine suitable circuit cable sizes, i.e. the minimum cross-sectional area for circuits.

The basis of cable 'size' selection includes consideration of the:
(i)   safe working temperature of the conductor;
(ii)   thermal effect on the cable's insulation;
(iii) effect of the heat produced on surrounding materials;
(iv)   thermal effect on the conductor's resistance; and
(v)   effects of voltage drop on the operation and the output rating of the connected load.

The cable conductor must dissipate the heat developed by the current flowing through it to the surrounding atmosphere, the electrical insulation around it and the cable enclosure, e.g. conduit, trunking or a building void may form heat insulation which will limit the cable's heat dissipation. Naturally the more cables run together the greater the heat developed. The ambient temperature which will exist in each area of the cable run when the circuit is working normally (i.e. at full load) must be established before any cable sizing calculation is attempted.

Volume 1 of this series of books introduced British Standard 7671:1992 and outlined the method of determining the cross-sectional area (c.s.a.) of circuit wiring cables in accordance with Appendix 4. It is worth repeating that section of Volume 1 before further demonstrating the method.

Reference to a copy of BS 7671:1992 will be advisable when attempting the exercises, but the reader must have at least a copy of the relevant cable current ratings and voltage drop tables, e.g. the *IEE On-site Guide*, before being able to tackle the exercises.

APPLICATION OF CORRECTION FACTORS

The cable current ratings (tabulated values) shown in Appendix 4 and in BS 7671 and the *IEE On-site Guide* represent the current which can be carried safely by the particular type of cable and stated installation method for a *single circuit* in an ambient temperature of 30 °C; any other combination of condition means that the cable must be rerated by the application of correction factors. More explanation will be given later in this section.

For practical purposes we will follow the same method of calculation as in Volume 1 of this series. The first step in calculating the minimum current-carrying capacity or current rating of a circuit cable is to establish the design current of the circuit (I_b), then select a suitably rated overcurrent protective device (I_n) in accordance with BS 7671 requirement 433-02.

The minimum tabulated cable current rating (I_t) is then determined by the use of the formula

$$I_t = \frac{I_n}{C_a C_i C_g C_r}$$

where:

I_n is the current rating of the fuse or the setting of the circuit breaker protecting the circuit

C_a is a correction factor for ambient temperature (Table 6A1 and 6A2 in the *IEE On-site Guide*)

C_i is a correction factor for thermal insulation (Table 6B in the *IEE On-site Guide*)

C_g is a correction factor for cable grouping (Table 6C in the *IEE On-site Guide*)

C_r is a correction factor used when semi-enclosed fuses to BS 3036 (rewireable fuses) are used.

Not all the factors may be necessary, as it is often possible to determine a cable rating based simply upon the most 'hostile' environmental area and then accept it as the minimum cable size to be used throughout the cable run.

The correction factor C_r (0.725) however *must* be used whenever semi-enclosed (rewirable) fuses protect a circuit wired in pvc-insulated cables; however, where mineral insulated cables are used then C_r may be disregarded.

To establish the actual current-carrying capacity of a cable (I_z) under normal operating conditions we may employ the formula

$$I_z = I_t \times C_a \times C_i \times C_g \times C_r$$

Take as an example a $1.5\,\text{mm}^2$ single-core pvc-insulated cable (copper conductors) with an I_t value of 20 A and correction factors of $C_a = 0.97$, $C_i = 1.0$, $C_g = 0.7$ and $C_r = 0.725$. Then

$$I_z = 20 \times 0.97 \times 1.0 \times 0.7 \times 0.725 = 9.85\,\text{A}$$

(approximately half the I_t value).

VOLTAGE DROP CALCULATIONS

The voltage drop in cable conductor(s) is directly proportional to the circuit current and the length of cable run.

Voltage drop

$$= \frac{\text{current (A)} \times \text{length of run (m)} \times \text{millivolt drop per A/m}}{1000}$$

(Note division by 1000 to convert millivolts to volts.)

Note BS 7671 Requirement 525-01-02 limits the voltage drop permitted between the origin of the installation and the terminals of a load to 4% of the nominal supply voltage. For a single-phase 230 V supply this equates to 9.2 V, and for 400 V three-phase supply to 16 V.

EXAMPLE I A 3 kW 230 V 50 Hz single-phase motor has an efficiency of 70% and works at a power factor of 0.6. It is connected to its starter by single-core pvc-insulated cables (copper conductors) drawn into steel conduit (method 3); the length of run is 25 m. The voltage drop in the cables must not exceed 6 V. Assume an ambient temperature of 35 °C and protection by BS 88 fuses.

Circuit details:
Motor circuit, starter will offer overload protection.
Ambient temperature 35 °C so C_a is 0.94.
Using BS 88 (Gm) fuses so C_r is 1.

$$\text{Output} = 3\,\text{kW}$$

$$\text{Input} = 3000 \times \frac{100}{70}$$

$$= 4285.7\,\text{W}$$

$$P = U \times I \times \text{p.f.}$$

$$4285.7 = 230 \times I \times 0.6$$

$$I_\text{b} = \frac{4285.7}{230 \times 0.6}$$

$$= 31.1\,\text{A}$$

Minimum BS 88 fuse rating (I_n) say 40 A (allows for moderate overcurrent at starting). Starter will offer overload protection (see BS 7671 requirements 435-01-01 and 552-01-02).

Correction factors applying:

C_a is 0.94 (35 °C)

C_r is 1 (BS 88 fuses)

Thus minimum current rating:

$$(I_\text{t}) = \frac{40}{0.94 \times 1}$$

$$= 42.55\,\text{A}$$

Using BS 7671 Table 4D1A or *IEE On-site Guide* Table 6D1, from column 4 select 10 mm^2 cables (57 A) and using BS 7671 Table 4D1B or *IEE On-site Guide* Table 6D2, column 3, read mV/A/m value for 10 mm^2 cables as 4.4 mV/A/m

$$\text{Volts drop in 25\,m} = \frac{31.1 \times 25 \times 4.4}{1000}$$

$$= 3.42\,\text{V}$$

Thus 10 mm^2 cables will be suitable.

EXAMPLE 2

(a) An industrial process heater of rating 16 kW is fed at 400 V 50 Hz. Three-phase four-wire is to be installed in a factory using a pvc-insulated, non-armoured, copper conductors multicore cable. Length of run is 25 m clipped direct to a wall; assume a maximum ambient temperature of 35 °C and protection by BS 3036 fuses.

(b) If the BS 3036 fuses were replaced by BS 88 (Gg) fuses what would be the effect on cable current rating?

Circuit details:

As it is a heater p.f. is unity.

Ambient temperature 35 °C so C_a is 0.94.

Using BS 3036 fuses so C_r is 0.725.

$$\text{Current demand } I_b = \frac{16\,000}{\sqrt{3} \times 400}$$

$$= 23.1\,\text{A}$$

(a) Select as I_n 30 A BS 3036 fuses. Thus minimum current rating is

$$I_t = \frac{30}{0.94 \times 0.725}$$

$$= 44\,\text{A}$$

Using BS 7671 Table 4D2A or *IEE On-site Guide* Table 6E1, from column 7 select 10 mm^2 cables (57 A) and using BS 7671 Table 4D2B or *IEE On-site Guide* Table 6E2, column 4, read mV/A/m value for 10 mm^2 cables as 3.8 mV/A/m.

$$\text{Volts drop in } 25\,\text{m} = \frac{23.1 \times 25 \times 3.8}{1000}$$

$$= 2.19\,\text{V}$$

Thus 10 mm^2 cables will be suitable.

(b) Select as I_n 25 A BS 88 fuses. In this case C_r is 1. Thus minimum current rating is

$$I_t = \frac{25}{0.94 \times 1}$$

$$= 26.6\,\text{A}$$

Using BS 7671 Table 4D2A or *IEE On-site Guide* Table 6E1, from column 7 select 4 mm^2 cables (32 A) and using BS 7671 Table 4D2B or *IEE On-site Guide* Table 6E2, column 4, read mV/A/m value for 4 mm^2 cables as 9.5 mV/A/m.

$$\text{Volts drop in } 25\,\text{m} = \frac{23.1 \times 25 \times 9.5}{1000}$$

$$= 5.49\,\text{V}$$

Thus $4\,\text{mm}^2$ cables will be suitable.

EXAMPLE 3 A pvc trunking containing single-core pvc-insulated distribution cables (copper conductors) is to be run 30 m from the main switchgear of an office building to supply a new 400/230 V TP & N distribution fuseboard. The balanced load consists of 18 kW of discharge lighting. The main and local distribution boards employ fuses to BS 88 (Gg) Part 2. The voltage drop in the distribution cables must not exceed 6 V and the ambient temperature is anticipated to be 30 °C.

For the distribution cables, establish and state the

(i)   design current I_b

(ii) minimum rating of fuse in the main switch fuse I_n

(iii) maximum mV/A/m value

(iv) minimum current rating I_t

(v)   minimum cross-sectional area of the live conductors

(vi) actual voltage drop in the cable   (C & G)

Circuit details:

Discharge lighting circuit requires a multiplier of 1.8 (*IEE On-site Guide*, Appendix 1).

Ambient temperature 30 °C (normal) so C_a is 1.

Using BS 88 fuses so C_r is 1.

Cable voltage drop limitation of 6 V.

Cables in trunking to method 3.

(i)   Design current $I_b = \dfrac{18 \times 10^3 \times 1.8}{\sqrt{3} \times 400}$

$$= 46.77\,\text{A}$$

(ii) Minimum BS 88 fuse rating is 50 A.

(iii) Maximum mV/A/m value $= \dfrac{6 \times 1000}{46.77 \times 30}$

$$= 4.28\ \text{mV/A/m}$$

(iv)   Minimum current rating $= \dfrac{50}{1}$

$\qquad\qquad\qquad\qquad = 50\,\text{A}$

(v)   From BS 7671 Tables 4D1A and 4D1B or *IEE On-site Guide* Tables 6D1 (column 5) and Table 6D2 (column 6), minimum c.s.a. of cable is $16\,\text{mm}^2$ (68 A/2.4 mV/A/m).

(vi)   Actual voltage drop in 30 m cable $= \dfrac{46.77 \times 30 \times 2.4}{1000}$

$\qquad\qquad\qquad\qquad\qquad\qquad = 3.37\,\text{V}$

EXAMPLE 4   A 400 V 50 Hz three-phase extract fan has a rating of 15 kW at 0.8 p.f. lagging and is supplied from a BS 88 (Gg) Part 2 type distribution board 40 m distant. The cables are to be single-core, pvc-insulated, run in steel trunking with three similar circuits. Assume an ambient temperature of 35 °C and that the voltage drop in the cables is limited to 2.5% of the line voltage.

Establish the:

(i)   full load current of the motor I_L
(ii)   rating of the fuses I_n
(iii) minimum current rating of cables
(iv)   minimum cable c.s.a.
(v)   actual voltage drop in cables

   Circuit details:
   Extract fan circuit: low starting current.
   Four sets of circuit cables: C_g is 0.65.
   Ambient temperature: 35 °C so C_a is 0.94.
   Using BS 88 fuses, so C_r is 1.
   Cable voltage drop limitation of 2.5% of 400, i.e. 10 V.
   Cables in trunking to method 3.

(i)   As $P = \sqrt{3}\,U_L I_L \cos\phi$

$\qquad 15\,000 = \sqrt{3} \times 400 \times I_L \times 0.8$

$\qquad\quad I_L = \dfrac{15\,000}{\sqrt{3} \times 400 \times 0.8}$

$\qquad\qquad = 27\,\text{A}$

(ii) Select 32 A BS 88 fuses (allowing for low starting current).

(iii) Minimum current rating of cables $= \dfrac{32}{0.94 \times 0.65}$

$$= 52.4\,\text{A}$$

(iv) From BS 7671 Tables 4D1A and 4D1B or *IEE On-site Guide* Tables 6D1 (column 5) and Table 6D2 (column 6) select $16\,\text{mm}^2$ (68 A/2.4 mV/A/m).

(v) Voltage drop in 40 m $= \dfrac{27 \times 40 \times 2.4}{1000}$

$$= 2.6\,\text{V}$$

As volts drop limitation is 4% of 400 V, i.e. 16 V, $16\,\text{mm}^2$ cable is satisfactory.

The final example illustrates the effect on the required tabulated cable rating of combined correction factors.

EXAMPLE 5   A twin and earth pvc-insulated (copper conductors) cable runs between a 230 V distribution board at the origin of an installation and a 10 kW heater. The cable passes through the following environmental conditions:

(a) on its own in a switchroom with an ambient temperature of 35 °C;

(b) on its own in an outdoor area with an ambient temperature of 25 °C;

(c) bunched with three other cables on a wall surface in an area with an ambient temperature of 40 °C;

(d) finally on its own passing through a thermally insulated wall section for a distance of 2 m, in an ambient temperature of 30 °C.

Protection is by BS 3036 fuses, length of run is 60 m and the voltage drop is limited to 5.5 V.

Calculate the minimum cable rating and select suitable cable for voltage drop limitation.

Circuit details:

Heater circuit so no special restrictions.

Protection by BS 3036 fuses so C_r is 0.725.

Voltage drop limitation is 5.5 V.

Area (a)   35 °C C_a is 0.94.
Area (b)   25 °C C_a is 1.03.
Area (c)   40 °C C_a is 0.87, C_g is 0.65.
Area (d)   30 °C C_a is 1, C_i is 0.5.

Now overall correction factors are as follows:

Area (a) $0.94 \times 0.725 = 0.68$
Area (b) $1.03 \times 0.725 = 0.747$
Area (c) $0.87 \times 0.65 \times 0.725 = 0.41$
Area (d) $0.5 \times 0.725 = 0.36$ (worst area)

$$\text{Design current } I_b = \frac{10\,000}{230}$$

$$= 43.5\,\text{A}$$

Nearest BS 3036 fuse element is 45 A (BS 7671 Table 53A). Select worst area (d): C_a is 0.5.

$$\text{Minimum cable rating} = \frac{45}{0.36}$$

$$= 125\,\text{A}$$

From BS 7671 Table 4D2A or *IEE On-site Guide* Table 6E1 and from BS 7671 Table 4D2B or *IEE On-site Guide* Table 6E2 select $70\,\text{mm}^2$ (139 A) and 0.63 mV/A/m.

$$\text{Volts drop in } 60\,\text{m} = \frac{43.5 \times 60 \times 0.63}{1000}$$

$$= 1.64\,\text{V}$$

So $70\,\text{mm}^2$ cable is satisfactory.

Obviously one should avoid running cables in hostile environments wherever possible, in this case avoiding thermal insulation and not using BS 3036 protection. Assuming that the cable grouping was unavoidable we could now use area (c) as the worst environment and in this case:

Revised circuit details:
Heater circuit so no special restrictions.
Protection by BS 88 fuses so C_r is 1.
Voltage drop limitation is 5.5 V.

Area (c) $0.87 \times 0.65 = 0.565$.

Voltage drop limitation 5.5 V.

$$\text{Minimum cable rating} = \frac{45}{0.565}$$
$$= 79.6\,\text{A}$$

From BS 7671 Table 4D2A or *IEE On-site Guide* Table 6E1 and from BS 7671 Table 4D2B or *IEE On-site Guide* Table 6E2 select $25\,\text{mm}^2$ (90 A) and 1.75 mV/A/m.

$$\text{Volts drop in 60 m} = \frac{43.5 \times 60 \times 1.75}{1000}$$
$$= 4.57\,\text{V}$$

So $25\,\text{mm}^2$ cable is satisfactory and cheaper to install than $70\,\text{mm}^2$ cable.

EXERCISE 16

1. Establish the current-carrying capacity (I_z) of a cable with a tabulated current rating (I_t) of 17.5 A when it is grouped in conduit with two other circuits in an ambient temperature of 35 °C; protection is by BS 3036 fuses.

2. Calculate the actual voltage drop and the power wasted in a $25\,\text{mm}^2$ cable, 10 m long, when it carries 70 A. The listed mV/A/m for the cable is 1.8 mV.

3. The design current of a single-phase circuit is 35 A. The single-core pvc-insulated cables run alone in pvc conduit for a distance of 50 m through an area having an ambient temperature of 35 °C (100 mm of the conduit passes through thermal insulation). The voltage drop in the circuit must not exceed 5 V. Protection is by a BS 1361 fuse. Determine the:
   (a) fuse rating
   (b) minimum cable current rating
   (c) minimum cable c.s.a.
   (d) voltage drop in the cables.

4. A supply is required to a 3 kW heater which is 25 m from a local BS 1361 distribution board. The building is fed at 230 V 50 Hz single-phase. It is proposed to employ a $2.5\,\text{mm}^2$ two-core and earth pvc-insulated (copper conductors) cable for this circuit installed as method 1. Allowing for a 2 V drop in the cables feeding the distribution board,

determine the:

(a) design current

(b) maximum volts drop permitted

(c) volts drop in the cable

(d) actual voltage at the heater.

5. A 10 kW motor having an efficiency of 60% is fed from a 220 V d.c. supply through cables 20 m long and having a listed voltage drop figure of 1.3 mV/A/m. Determine the:

(a) design current

(b) volts drop in the cables when the motor is fully loaded.

6. After the application of correction factors, a pair of single-core pvc-insulated cables in conduit are required to carry 25 A from a distribution board to a load 90 m away. The voltage drop in the cables should not exceed 5 V. Using BS 7671 documents:

(a) calculate the maximum mV/A/m value

(b) select a suitable cable c.s.a.

(c) calculate the voltage drop in the cables.

7. A 12 kW load is to be supplied from a 230 V main switch-fuse 65 m distant. The voltage drop is to be limited to 2.5% of the supply voltage. Overload protection is to be provided by a BS 3036 semi-enclosed fuse. The single-core pvc-insulated cables run in conduit with one other single-phase circuit. Assuming an ambient temperature of 25 °C, determine with the aid of BS 7671 documents:

(a) the design current

(b) the fuse rating

(c) the minimum cable current rating

(d) the maximum mV/A/m value

(e) the selection of a suitable cable c.s.a.

(f) the voltage drop in the circuit.

8. A single-phase load of 10 kW is to be supplied from a 230 V distribution board 120 m distant. Overload protection is to be by BS 88 (Gg) Part 2 fuses. The twin with earth, pvc-insulated cable is clipped with three similar cables as BS 7671 method 1 in an ambient temperature of 25 °C. Voltage drop in the cables should not exceed 5 V. Determine with the aid of BS 7671 documents the:

(a) design current

(b) fuse rating

(c) minimum cable current rating

(d) maximum mV/A/m value

(e) minimum cable c.s.a.

(f) voltage drop in the cables.

9. A 400/230 V 50 Hz TP & N distribution board is to be installed in a factory to feed 11 kW of mercury vapour lighting. Due to the adverse environmental conditions, it is intended to use pvc conduit to contain the single-core pvc-insulated cables (copper conductors). The total length of the run from the main switchboard is 50 m. To provide earthing protection it is intended to draw a 4 mm^2 single-core pvc-insulated cable (copper conductors) into the conduit. The following details apply to the installation:

(i) an ambient temperature of 35 °C

(ii) BS 88 (Gg) Part 2 fuse protection throughout

(iii) voltage drop in the cables must not exceed 8.5 V

(iv) the BS 88 fuses satisfy the requirements of BS 7671 Requirement 434-03-03.

Establish the:

(a) design current

(b) rating of fuses in the main switchboard

(c) minimum current rating of live conductors

(d) maximum mV/A/m value of live conductors

(e) minimum cross-sectional area of live conductors

(f) actual voltage drop in submain cables

(g) size of pvc conduit to be used, assuming that one section of the run involves one right-angle bend in 8 m.   (C&G)

10. A single steel trunking is to be run from a 400/230 V 50 Hz main switchboard to feed three SP & N lighting distribution boards containing Type 2 BS 3871 miniature circuit breakers, sited at 5, 12 and 20 m distances. Each distribution board feeds 5 kW of mercury vapour lighting. The following details apply to the installation:

(i) ambient temperature in the area is 25 °C

(ii) protection at the main switchboard is by BS 88 fuses

(iii) single-core pvc-insulated (copper conductors) cables are to be employed

(iv) voltage drop in the distribution cables must not be greater than 3.5 V.

Establish the:

(a) design current

(b) maximum mV/A/m value permitted

(c) fuse rating at the main switchboard

(d) minimum cable current rating

(e) minimum cross-sectional area of the distribution cables

(f) voltage at each distribution board.   (C&G)

11. A 12 m length of two-core and earth, pvc-insulated cable is clipped to a surface as BS 7671 method 1. The cable feeds a load of 4 kW at 230 V 50 Hz a.c. The following details apply to the installation:

(i) power factor of the load is 0.8 lagging

(ii) ambient temperature of 30 °C

(iii) protection by a BS 88 (Gg) Part 2 fuse

(iv) cable voltage drop not to exceed 5 V.

Determine the:

(a) design current

(b) rating of the fuse

(c) minimum cable current rating

(d) maximum mV/A/m value

(e) minimum cable cross-sectional area

(f) actual voltage drop in the cable at full load.

12. A 4.5 kW single-phase load in a factory is to be supplied from the 400/230 V 50 Hz suppliers' main switchboard 40 m distant, using two-core and earth, pvc-insulated cable. The power factor of the load is 0.7 lagging and the cable route is through an ambient temperature of 30 °C. Protection is by BS 88 (Gg) Part 2 fuses.

Determine the:

(a) design current

(b) permissible voltage drop in circuit

(c) minimum fuse rating

(d) minimum cable current rating

(e) maximum mV/A/m value

(f) minimum cable cross-sectional area

(g) actual voltage drop in the cable at full load.

13. A 230 V 50 Hz 8 kW electric shower unit is to be installed in an industrial premises using a two-core and earth, pvc-insulated cable, 20 m in length. The ambient temperature is 30 °C. Protection is by a BS 1361 fuse in a

distribution board, the cable volts drop should not exceed 2 V.

Determine the:

(a) design current I_b

(b) fuse rating I_n

(c) required cable current rating I_t

(d) required cable c.s.a.

(e) actual volts drop in the cable.

14. A 25 kW 400 V 50 Hz three-phase motor operates at 0.85 p.f. lagging on full load. The pvc-insulated single-core cables run together for a distance of 10 m with two similar circuits through a trunking to a circuit breaker distribution board. Assume that the circuit breaker is selected to have an operating value of not less than 1.5 times the motor full-load current, the ambient temperature is 35 °C and the voltage drop in the cables should not exceed 10 V. Determine the:

(a) design current

(b) setting of circuit breaker

(c) minimum cable current rating

(d) maximum mV/A/m value

(e) minimum cable c.s.a.

(f) actual volts drop in the cable.

15. The voltage-drop figure for a certain cable is 2.8 mV/A/m. The actual drop in 50 m run of this cable when carrying 45 A is:

(a) 1.2 V   (b) 6.3 V   (c) 0.1 V   (d) 10 V

16. The voltage drop allowed in a certain circuit is 6 V. The length of run is 35 m. The cable used has a voltage-drop figure of 7.3 mV/A/m. Ignoring any correction factors, the maximum current which the cable can carry is:

(a) 15 A   (b) 23.5 A   (c) 41 A   (d) 43.8 A

17. A circuit is given overload protection by a 30 A BS 3036 fuse. The grouping factor C_g is 0.65 and the ambient temperature factor is 0.87. The minimum current-carrying capacity of the cable should be:

(a) 73.2 A   (b) 53 A   (c) 30 A   (d) 41.3 A

18. A 10 kW 230 V a.c. motor operates at 0.75 lagging. The starter offering overload protection is set at 1.5 times the F.L.C. of the motor. Ignoring any correction factors, the minimum current-carrying capacity of the cable to the motor required is:

(a) 43.5 A   (b) 58 A   (c) 87 A   (d) 108.7 A

19. A certain cable having a tabulated current rating (I_t) of 18 A has correction factors of 1.04, 0.79 and 0.725 applied to compensate for its operating conditions. The operational current rating (I_z) for the cable is:

(a) 30.22 A **(b)** 24.83 A **(c)** 13.572 A **(d)** 10.72 A

EARTH LEAKAGE PROTECTION CALCULATIONS

To prevent danger to persons, livestock and property every installation must be protected against the persistence of earth leakage currents. This is generally achieved by providing a low-impedance earth-current leakage path from the installation to the source of supply, i.e. the local distribution transformer.

The leakage path must have a low enough impedance (Z_s) to allow adequate earth leakage current to flow to 'blow' the circuit fuse or operate the circuit breaker and isolate the faulty circuit within a specified time, usually either 5 seconds or 0.4 seconds. BS 7671 gives guidance to the permissible earth-loop impedance values to meet the disconnection times and that document and the *IEE On-site Guide* contain tables which list types of protective device and specify the maximum measured earth fault loop impedance in ohms for each rating of the specific device. Where precise disconnection times are demanded then BS 7671 Appendix 3 contains characteristic curves for fuses and circuit breakers.

Part of the earth leakage path is outside the control of an electricity consumer and its impedance (Z_e) contributes to the total value of earth loop impedance. The value of this external impedance is generally declared by the supplier and is used in the calculation of the 'prospective' Z_s. The declared value of Z_e, however, can never be a precise value because of the supplier's service conditions at the moment of earth fault; thus the actual value of Z_s must always be measured by earth loop impedance test instruments at various points within an installation when the particular circuit is energized and is under test-load conditions.

For the estimation of prospective earth-loop impedance values we may however regard Z_e as an empirical or estimated value when assessing the permitted value of the installation's internal cable impedance (or resistance) value.

The internal cable 'impedance' will be determined by the cross-sectional area and resistance (R_1) of the circuit's phase conductor and that of the circuit's protective conductor (R_2) from the origin of the installation to the point of connection to current-using equipment when the circuit is energized and the cables are working in their maximum operating temperature.

To predict the actual disconnection time for an earth leakage fault condition we may employ characteristic curves of the protective devices, i.e. fuses and circuit breakers. Appendix 3 of BS 7671 gives specimens of such curves.

Note For all the following examples and exercises pvc-insulated copper conductors are to be employed.

EXAMPLE I An installation is being carried out and it is necessary to estimate the prospective total earth loop impedance of circuits. In order to arrive at a typical value, a lighting circuit is chosen as that is likely to have a fairly high impedance value. The circuit is to be wired in $1.5\,\text{mm}^2$ twin and earth cable (assume a $1.0\,\text{mm}^2$ protective conductor); the length of cable is 18 m. The declared value of Z_e is $0.35\,\Omega$. Circuit protection at the origin of the installation (consumer unit) is by a BS 1361 5 A fuse.

(a) Establish conformity with BS 7671 requirements.

(b) Establish from BS 7671 Appendix 3, the actual disconnection time.

This is a fixed-equipment circuit; five-second disconnection time.

From Tables 9A and 9C (*IEE On-site Guide*) $R_1 + R_2$ of $1.5\,\text{mm}^2/1.0\,\text{mm}^2$ conductors is $30.2\,\text{m}\Omega/\text{m} \times 1.20$.

Thus $R_1 + R_2$ of $1.5\,\text{mm}^2/1.0\,\text{mm}^2$ conductors 18 m long will be

$$\frac{18 \times 30.2 \times 1.20}{1000} = 0.65\,\Omega$$

and
$$Z_s = 0.35 + 0.65$$
$$= 1.0\,\Omega$$

(a) From Tables 2C and 2E (*IEE On-site Guide*) maximum measured earth fault loop impedance is $13.68 \times 1.06\,\Omega$, i.e. $14.5\,\Omega$, thus the estimated value of the earth fault loop impedance for this circuit is acceptable.

(b) Actual disconnection time

$$\text{Prospective earth fault current} = \frac{230}{1}$$

$$= 230\,\text{A}$$

From Appendix 3, Table 3.1, the circuit disconnects in less than 0.1 second; we may say that the fuse operates instantaneously.

EXAMPLE 2 A commercial cooker circuit is fed by $16\,\text{mm}^2$ single-core pvc-insulated cable with a $6\,\text{mm}^2$ single-core pvc-insulated protective conductor cable from a BS 88 (Gg) Part 2 type fuseboard (40 A fuse) at the origin of the installation; length of cables within pvc conduit is 35 m. Assume a tested Z_e value of $0.7\,\Omega$.

(a) Establish conformity with BS 7671 requirements regarding the value of Z_s.

(b) Establish from BS 7671 Appendix 3 the actual disconnection time.

(a) This is a fixed-equipment circuit; five-second disconnection time. From Tables 9A and 9B (*IEE On-site Guide*) $R_1 + R_2$ of $16\,\text{mm}^2/6\,\text{mm}^2$ conductors is $4.23\,\text{m}\Omega/\text{m} \times 1.20$.

Thus $R_1 + R_2$ of $16\,\text{mm}^2/6\,\text{mm}^2$ conductors 15 m long will be

$$\frac{35 \times 4.23 \times 1.20}{1000} = 0.178\,\Omega$$

and
$$Z_s = 0.7 + 0.178$$
$$= 0.878\,\Omega$$

From Tables 2B and 2E (*IEE On-site Guide*) maximum measured earth fault loop impedance is $1.13 \times 1.06\,\Omega$, i.e. $1.2\,\Omega$, thus the estimated value of the earth fault loop impedance for this circuit is acceptable.

(b) Actual disconnection time

$$\text{Prospective earth fault current} = \frac{230}{0.878}$$

$$= 261\,\text{A}$$

From Appendix 3, Table 3.3B, the circuit disconnects in 0.6 seconds.

EXAMPLE 3 A cable feeds a single-phase electric pump and lighting point in an external building of a factory, the total length of the twin with earth $4\,\text{mm}^2/2.5\,\text{mm}^2$ pvc-insulated cable is 30 m, protection is by a BS 3871, type 3, 30 A mcb in a distribution board at the suppliers' intake position. The tested value of Z_s at the intake position is $0.35\,\Omega$.

This is a fixed-equipment circuit but is in adverse conditions, thus a 0.4 s disconnection time applies (BS 7671 requirement 471-08-03).

From Tables 9A and 9B (*IEE On-site Guide*) $R_1 + R_2$ of $4\,\text{mm}^2/2.5\,\text{mm}^2$ conductors is $12.02\,\text{m}\Omega/\text{m} \times 1.20$.

Thus $R_1 + R_2$ of $4\,\text{mm}^2/2.5\,\text{mm}^2$ conductors 30 m long will be

$$\frac{30 \times 12.02 \times 1.20}{1000} = 0.43\,\Omega$$

and
$$Z_s = 0.35 + 0.43$$
$$= 0.78\,\Omega$$

From Tables 2D and 2E (*IEE On-site Guide*) maximum measured earth fault loop impedance is $0.64 \times 1.06\,\Omega$, i.e. $0.67\,\Omega$, thus the estimated value of the earth fault loop impedance for this circuit is not acceptable. A residual current device in the supply to the external building will be necessary.

EXERCISE 17

Note Assume that copper conductor cables are used for all questions.

1. A single-phase process heater circuit is wired in pvc trunking and is to employ $6\,\text{mm}^2$ single-core pvc-insulated live conductors and a $2.5\,\text{mm}^2$

protective conductor. The distance from the BS 88 (Gg) Part 2 distribution fuseboard at the main switchgear is 33 m, rating of fuse is 40 A and ambient temperature is 25 °C. The tested Z_e value at the main switchgear is 0.3 Ω.

   (a) Estimate the prospective value of Z_s.

   (b) State the maximum permissible measured Z_s value.

2. A single-phase lighting circuit in a commercial premises is wired in pvc conduit employing 1.5 mm^2 live conductors and a 1.5 mm^2 protective conductor. The distance from the BS 1361 distribution fuseboard at the main switchgear is 20 m, rating of the fuse is 15 A and ambient temperature is 30 °C. The tested Z_e value at the main switchgear is 0.45 Ω.

   (a) Estimate the prospective value of Z_s.

   (b) State the maximum permissible measured Z_s value.

3. A three-phase electric motor circuit is to be wired in steel trunking and is to employ 4 mm^2 live conductors and the client demands that an independent 2.5 mm^2 protective conductor is used. The distance from the BS 88 (Gm) Part 2, distribution fuseboard at the main switchgear is 10 m, and the rating of the fuse is 10 A. The tested Z_s value at the distribution board is 0.45 Ω, and the ambient temperature is 25 °C.

   (a) Estimate the prospective value of Z_s at the motor starter.

   (b) State the maximum permissible measured Z_s value.

4. A 400/230 V 50 Hz three-phase milling machine is to be wired in pvc trunking and is to employ 6 mm^2 live conductors and the designer specifies that an independent 4 mm^2 protective conductor is to be used. The distance from the BS 88 (Gm) Part 2, distribution fuseboard at the main factory switchgear is 18 m, and the rating of the fuse is 50 A. The tested Z_s value at the distribution board is 0.4 Ω, and the ambient temperature is 20 °C.

   (a) Estimate the value of Z_s at the machine's starter isolator.

   (b) Assuming that the actual value of Z_s is as estimated, what earth fault current will flow in the event of a direct to earth fault at the isolator?

   (c) What will be the approximate disconnection time?

5. An earth fault current of 250 A occurs in a circuit protected by a BS 88 (Gg) Part 2 32 A fuse. The disconnection time will be approximately:

   (a) 0.1 s   **(b)** 0.2 s   **(c)** 0.25 s   **(d)** 3 s

6. An earth fault current of 130 A occurs in a circuit protected by a BS 3036 30 A fuse. The disconnection time will be approximately:

   (a) 0.8 s   **(b)** 0.13 s   **(c)** 1 s   **(d)** 8 s

7. An earth fault current of 300 A occurs in a circuit protected by a BS 1361 45 A fuse. The disconnection time will be approximately:

(a) 0.18 s   **(b)** 1.8 s   **(c)** 0.3 s   **(d)** 0.9 s

Mechanics

MOMENT OF FORCE

The moment of force about a point is found by multiplying together the force and the perpendicular distance between the point and the line of action of the force.

Consider an arm attached to a shaft as in Fig. 78. The moment acting on the shaft tending to turn it clockwise is

$$2\,\text{N} \times 0.5\,\text{m} = 1\,\text{Nm}$$

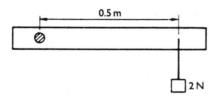

Fig. 78

TORQUE

If in Fig. 78 a turning effect is applied to the shaft in the opposite direction so that the arm is maintained in a horizontal position, then the *torque* exerted at the shaft is 1 Nm.

Consider now an electric motor fitted with a pulley 0.25 m in diameter over which a belt passes to drive a machine (Fig. 79). If the pull on the tight side of the belt is 60 N when the motor is running,

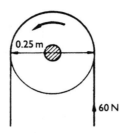

Fig. 79

then a continuous torque of

$$60\,\text{N} \times \frac{0.25\,\text{m}}{2} = 7.5\,\text{Nm is present}$$

This ignores any pull on the slack side of the belt, and this must usually be taken into account. Thus if the tension in the slack side of the belt is say 10 N, then the net torque exerted by the motor is

$$(60 - 10)\,\text{N} \times \frac{0.25\,\text{m}}{2} = \frac{50 \times 0.25\,\text{Nm}}{2}$$

$$= 6.25\,\text{Nm}$$

In general the torque exerted is

$$T = (F_1 - F_2) \times r\,\text{Nm}$$

where F_1 is the tension in the tight side, F_2 is the tension in the slack side (in newtons), and r is the pulley radius (in metres).

POWER

$$P = 2\pi n T \text{ watts}$$

where T is the torque in newton metres and n is the speed of the pulley in revolutions per second.

EXAMPLE I If the pulley previously considered is running at 16 rev/s, calculate the power output of the motor.

$$P = 2\pi n T$$
$$= 2\pi \times 16 \times 6.25$$
$$= 629\,\text{W}$$

EXAMPLE 2   Calculate the full-load torque of a 3 kW motor running at 1200 rev/min.

$$1200\,\text{rev/min} = \frac{1200}{60} = 20\,\text{rev/s}$$

$$P = 2\pi n T$$

$$\therefore \quad 3 \times 1000 = 2\pi \times 20 \times T \quad \text{(note conversion of kW to W)}$$

$$\therefore \quad\quad T = \frac{3 \times 1000}{2\pi \times 20}$$

$$= 23.9\,\text{Nm}$$

EXAMPLE 3   During a turning operation, a lathe tool exerts a tangential force of 700 N on the 100 mm diameter workpiece.
(a) Calculate the power involved when the work is rotating at 80 rev/min.
(b) Calculate the current taken by the 230 V single-phase a.c. motor, assuming that the lathe gear is 60% efficient, the motor is 75% efficient, and its power factor is 0.7.
The arrangement is shown in Fig. 80.
(a) The torque exerted in rotating the work against the tool is

$$T = 700\,\text{N} \times 0.05\,\text{m} \quad \text{(note: radius is 50 mm} = 0.05\,\text{m)}$$

$$= 35\,\text{Nm}$$

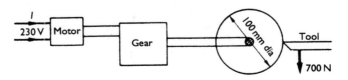

Fig. 80

$$P = 2\pi nT$$

$$= \frac{2\pi \times 80 \times 35}{60} \quad \text{(note: conversion of rev/min to rev/s)}$$

$$= 293 \, \text{W}$$

(b) $\text{Motor output} = 293 \times \dfrac{100}{60}$

$$= 488 \, \text{W}$$

$$\text{Motor input} = 488 \times \frac{100}{75}$$

$$= 650.6 \, \text{W}$$

$$P = V \times I \times \text{p.f.}$$

$\therefore$ $\quad\quad\quad 650.6 = 230 \times I \times 0.7$

$\therefore$ $\quad$ $\text{motor current } I = \dfrac{650.6}{230 \times 0.7}$

$$= 4.04 \, \text{A}$$

SURFACE SPEED, PULLEY DIAMETER AND SPEED RATIOS

EXAMPLE I   When turning a piece of low-carbon steel, it should be rotated so that the speed of its surface in relation to the tool is about 0.35 m/s. Determine the speed at which a bar 120 mm in diameter should be rotated in order to achieve this surface speed. Consider a point on the surface of the steel (Fig. 81). In one revolution, this point moves through a distance equal to the circumference of the bar.

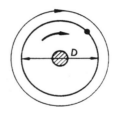

Fig. 81

i.e.   distance moved in one revolution $= \pi \times D$

$$= 3.142 \times \frac{120}{1000}$$

$$= 0.377 \,\text{m}$$

Number of revolutions $= \dfrac{0.35}{0.377}$
required for 0.35 m

$$= 0.9285$$

$\therefore$ speed of rotation $= 0.928 \,\text{rev/s}$

EXAMPLE 2   A machine is driven at 6 rev/s by a belt from a standard motor running at 24 rev/s. The motor is fitted with a 200 mm diameter pulley. Find the size of the machine pulley.

The speeds at which the pulleys rotate are inversely proportional to their diameters. Thus if the pulley having a diameter of D_1 rotates at n_1 rev/min and the pulley having a diameter of D_2 rotates at n_2 rev/min (Fig. 82), then

$$\frac{n_1}{n_2} = \frac{D_2}{D_1}$$

In this case,

$$\frac{24}{6} = \frac{D_2}{200}$$

$\therefore$ $D_2 = \dfrac{200 \times 24}{6}$

$$= 800 \,\text{mm}$$

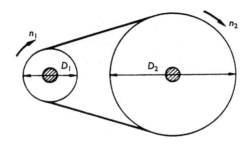

Fig. 82

1. A motor drives a machine by means of a belt. The tension in the tight side of the belt is 100 N, that in the slack side is 40 N, and the pulley is 200 mm in diameter. Calculate the total torque exerted by the motor.

2. A test on an induction motor fitted with a prony brake yielded the following results:

Tension in tight side of belt (N)	0	20	30	40	50	60
Tension in slack side of belt (N)	0	4	6	8.75	11.5	14.5
Speed (rev/min)	1450	1440	1430	1410	1380	1360

Calculate the torque and power corresponding to each set of readings. Take the pulley radius as being 100 mm.

3. A 10 kW motor fitted with a 250 mm diameter pulley runs at 16 rev/s. Calculate the tension in the tight side of the belt. Ignore any tension in the slack side.

4. A 4 kW motor fitted with a 150 mm diameter pulley runs at 24 rev/s. The tension in the tight side of the belt may be assumed to be equal to three times the tension in the slack side. Determine the tension in each side of the belt at full load.

5. Calculate the full-load torque of each of the motors to which the following particulars refer:

	Rated power (kW)	Normal speed (rev/min)
(a)	10	850
(b)	2	1475
(c)	18	750
(d)	0.25	1480
(e)	4	1200

6. A motor exerts a torque of 25 Nm at 16 rev/s. Assuming that it is 72% efficient, calculate the current it takes from a 440 V d.c. supply.

7. A brake test on a small d.c. motor, pulley diameter 75 mm, gave the following results:

Net brake tension (N)	0	5	10	15	20	25
Speed (rev/min)	1700	1690	1680	1670	1650	1640
Current (A)	0.8	1.05	1.3	1.68	1.9	2.25
Supply voltage (V)	116	116	116	116	116	116

For each set of values, calculate the power output and the efficiency. Plot a graph of efficiency against power.

8. The chuck of a lathe is driven at 2 rev/s through a gear which is 60% efficient from a 240 V d.c. motor. During the turning operation on a 75 mm diameter workpiece, the force on the tool is 300 N. Calculate the current taken by the motor, assuming its efficiency is 70%.

9. Calculate the speed at the circumference of a 250 mm diameter pulley when it is rotating at 11 rev/s.

10. A motor drives a machine through a vee belt. The motor pulley is 120 mm in diameter. Calculate the speed at which the belt travels when the motor runs at 24 rev/s.

11. The recommended surface speed for a certain type of grinding wheel is about 20 m/s. Determine the speed at which a 250 mm diameter wheel must rotate in order to reach this speed.

12. For a certain type of metal, a cutting speed of 0.6 m/s is found to be suitable. Calculate the most suitable speed, in revolutions per minute, at which to rotate bars of the metal having the following diameters in order to achieve this surface speed: **(a)** 50 mm, **(b)** 125 mm, **(c)** 150 mm, **(d)** 200 mm, **(e)** 75 mm.

13. A circular saw is to be driven at 60 rev/s. The motor is a standard one which runs at 1420 rev/min and is fitted with a 200 mm diameter pulley. Determine the most suitable size pulley for driving the saw.

14. **(a)** Calculate the speed of the smaller pulley in Fig. 83(a).
    (b) Determine the speed, in rev/min, of the larger pulley in Fig. 83(b).

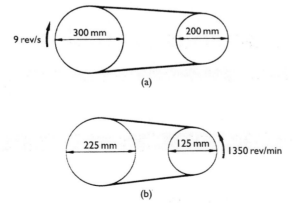

(a)

(b)

Fig. 83

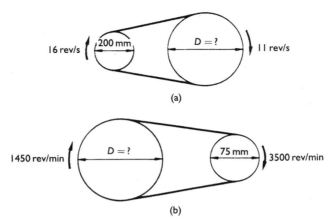

(a)

(b)

Fig. 84

15. Calculate the diameter of the larger pulley in Figs 84(a) and (b).

16. A cutting tool exerts a tangential force of 300 N on a steel bar 100 mm in diameter which is rotating at 160 rev/min in a lathe. The efficiency of the lathe gearing is 62% and that of the 240 V a.c. driving motor is 78%. Calculate the current taken by the motor if its power factor is 0.6.

The pulley on the lathe which takes the drive from the motor is 225 mm in diameter and rotates at 600 rev/min. The motor runs at 1420 rev/min. What is the diameter of the motor pulley?

Miscellaneous examples

D.C. GENERATORS

$$U = E - I_a R_a$$

where U is the terminal voltage,

   E is the generated e.m.f.,

I_a　is the armature current,

and　　　R_a　is the armature resistance.

EXAMPLE　Calculate the e.m.f. generated by a shunt generator which is delivering 15 A at a terminal voltage of 440 V. The armature circuit resistance is 0.15 Ω, the resistance of the shunt field is 300 Ω, and a voltage drop of 2 V occurs at the brushes.

The circuit is shown in Fig. 85.

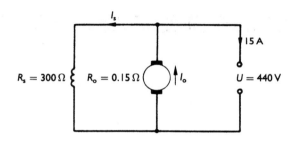

Fig. 85

To find the shunt field current,

$$U = I_s \times R_s$$

where　　　I_s　is shunt field current,

and　　　R_s　is shunt field resistance.

$\therefore$　　　$440 = I_s \times 300$

$\therefore$　　　　$I_s = \dfrac{440}{300}$

　　　　　$= 1.47 \, \text{A}$

Total armature current $= 15 + 1.47$

　　　　　　　　$= 16.47 \, \text{A}$

Neglecting the voltage drop at the brushes,

$$U = E - I_a R_a$$

$\therefore$　　　$440 = E - 16.47 \times 0.15$

　　　　　$= E - 2.47$

$$\therefore \quad 440 + 2.47 = E$$

$$\therefore \quad E = 442.47\,\text{V}$$

Allowing for the voltage drop at the brushes,

$$\text{generated e.m.f.} = 442.47 + 2$$

$$= 444.47$$

$$= 444\,\text{V}$$

D.C. MOTORS

$$U = E + I_{\text{a}}R_{\text{a}}$$

where U is the terminal voltage,

          E is the back e.m.f.,

          I_{a} is the armature current,

and R_{a} is the armature circuit resistance.

EXAMPLE   Calculate the back e.m.f. of a d.c. motor which is taking an armature current of 25 A from a 220 V supply. The resistance of its armature is 0.2 Ω.

$$U = E + I_{\text{a}}R_{\text{a}}$$

$$\therefore \quad 220 = E + 25 \times 0.2$$

$$= E + 5.0$$

$$\therefore \quad 220 - 5 = E$$

$$\therefore \quad E = 215\,\text{V}$$

ALTERNATORS AND SYNCHRONOUS MOTORS

$$f = n \times p$$

where f is the frequency in hertz,

          n is the speed in revolutions per second,

and p is the number of *pairs* of poles.

EXAMPLE I Calculate the number of poles in an alternator which generates 60 Hz at a speed of 5 rev/s.

$$f = n \times p$$

$$\therefore \quad 60 = 5 \times p$$

$$\therefore \quad p = \frac{60}{5}$$

$$= 12$$

$\therefore$ the machine has $2 \times p = 24$ poles

EXAMPLE 2 Calculate the speed at which a four-pole synchronous motor will run from a 50 Hz supply.

$$f = n \times p$$

$$\therefore \quad 50 = n \times 2 \quad \text{(4 poles gives 2 pairs)}$$

$$\therefore \quad n = \frac{50}{2}$$

$$= 25 \, \text{rev/s}$$

INDUCTION MOTORS

$$\text{Percentage slip} = \frac{n_s - n_r}{n_s} \times 100\%$$

where n_s is the synchronous speed,
and n_r is the actual speed of the rotor.

The synchronous speed n_s can be determined from the relationship

$$f = n_s \times p$$

as in the case of the synchronous motor.

EXAMPLE Calculate the actual speed of a six-pole cage-induction motor operating from a 50 Hz supply with 7% slip.

$$f = n_s \times p$$

$$\therefore \quad 50 = n_s \times 3$$

$$\therefore \quad n_s = \frac{50}{3}$$

$$= 16.7\,\mathrm{rev/s}$$

$$\text{Percentage slip} = \frac{n_s - n_r}{n_s} \times 100$$

$$\therefore \qquad 7 = \frac{16.7 - n_r}{16.7} \times 100$$

$$0.07 = \frac{16.7 - n_r}{16.7}$$

$$\therefore \qquad 0.07 \times 16.7 = 16.7 - n_r$$

$$\therefore \qquad n_r = 16.7 - 0.07 \times 16.7$$

$$= 15.5\,\mathrm{rev/s}$$

INSULATION RESISTANCE

The insulation resistance of a cable is *inversely* proportional to its length.

EXAMPLE 1 The insulation resistance measured between the cores of a certain twin cable 100 m long is 1000 MΩ. Calculate the insulation resistance of 35 m of the same cable.

The *shorter* length will have a *higher* value of insulation resistance because the path for the leakage current has less cross-sectional area (Fig. 86).

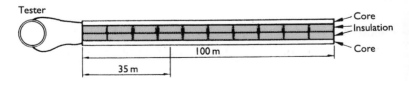

Fig. 86

Insulation resistance of $100 \, \text{m} = 1000 \, \text{M}\Omega$

$\therefore$ insulation resistance of $35 \, \text{m} = 1000 \times \dfrac{100 \, (larger)}{35 \, (smaller)}$

$$= 2857 \, \text{M}\Omega$$

EXAMPLE 2 The insulation resistance measured between the cores of a certain twin cable is $850 \, \text{M}\Omega$. Calculate the insulation resistance obtained when two such cables are connected (a) in series, (b) in parallel.

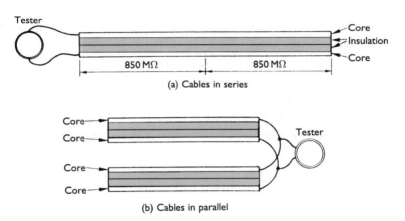

(a) Cables in series

(b) Cables in parallel

Fig. 87

It is seen from Fig. 87 that the effect in both cases is the same, i.e. to increase the c.s.a. of the leakage-current path through the insulation. The insulation resistance in either case is thus

$$\frac{850 \, \text{M}\Omega}{2} = 425 \, \text{M}\Omega$$

EXERCISE 19

1. What is meant by the expression 'back e.m.f.' of a direct-current motor? In what way does the back e.m.f. affect the starting of a direct-current motor?

   A direct-current motor connected to a 460 V supply takes an armature current of 120 A on full load. If the armature circuit has a resistance of 0.25 Ω, calculate the value of the back e.m.f. at this load. (CGLI)

2. A d.c. machine has an armature resistance of 8 Ω. Calculate

   (a) the back e.m.f. when it is operating from a 110 V supply and taking an armature current of 2.5 A;

   (b) the e.m.f. generated when the machine is running as a generator and delivering 2 A at a terminal voltage of 110 V. (Neglect the field current.)

3. A d.c. motor connected to a 460 V supply has armature resistance of 0.15 Ω. Calculate

   (a) the value of the back e.m.f. when the armature current is 120 A,

   (b) the value of the armature current when the back e.m.f. is 447.4 V.   (CGLI)

4. Explain briefly, with the aid of diagrams, the differences between series, shunt, and compound d.c. generators.

   A d.c. shunt generator delivers a current of 96 A at 240 V. The armature resistance is 0.15 Ω, and the field winding has a resistance of 60 Ω. Assuming a brush contact drop of 2 V, calculate (a) the current in the armature, (b) the generated e.m.f.   (CGLI)

5. Calculate the speed at which an eight-pole alternator must be driven in order to generate 50 Hz.

6. Calculate the frequency of the voltage generated by a four-pole alternator when it is running at (a) 16 rev/s, (b) 12 rev/s.

7. Determine the speed at which a six-pole synchronous motor will run from the 50 Hz mains.

8. The synchronous speed of an induction motor is 750 rev/min. The motor actually runs at 715 rev/min. Calculate the percentage slip.

9. A four-pole induction motor is operating at 24 rev/s from a 50 Hz supply. Calculate the percentage slip.

10. A cage-induction motor having six poles operates with a 4.5% slip from a 50 Hz supply. Calculate the actual rotor speed.

11. Calculate the full-load torque of a 30 kW six-pole 50 Hz induction motor, assuming that the slip at full load amounts to 5%.

12. Explain the term 'insulation resistance'. Describe, with wiring diagram, a suitable instrument for measuring insulation resistance.

    Calculate the insulation resistance of a 100 m coil of insulated cable. The insulation resistance of 1 km of the same cable is given as 2500 MΩ.

13. The insulation resistance of 1000 m of two-core cable is 1500 MΩ. Calculate the insulation resistance of

(a) 100 m **(b)** 200 m **(c)** 400 m **(d)** 600 m **(e)** 800 m

and plot a graph showing the relationship between cable length and insulation resistance.

14. Explain the term 'insulation resistance of an installation'. Describe, with connection diagram, the working of an instrument suitable for measuring insulation resistance.

    Three separate circuits are disconnected from a distribution board and tested for insulation resistance to earth. The respective values are 40 MΩ, 60 MΩ, and 300 MΩ. What is the combined insulation resistance to earth? (CGLI)

15. The insulation resistance measured between the cores of a certain twin cable is 950 Ω. Calculate the insulation resistance of three identical cables connected in parallel.

16. The resistance of an armature circuit of a motor is 1.2 Ω. The current through it is 15 A and the terminal voltage is 200 V. The generated e.m.f. is

    (a) 218 V **(b)** 182 V **(c)** 13.3 V **(d)** 125 V

17. An alternator generates 400 Hz at a speed of 500 rev/min. The number of pairs of poles is

    (a) 12 **(b)** 48 **(c)** 6 **(d)** 3

18. The insulation resistance measured between the cores of a cable is 900 MΩ for a 500 m length. The insulation resistance for 350 m of this cable would be:

    (a) 1285.7 MΩ **(b)** 630 MΩ **(c)** 157.5×10^6 MΩ **(d)** 194.4 MΩ

Answers

Exercise 1

1.
Volts V (a.c.)	10	225	230	400	100	25	230	625
Current (A)	0.1	15	0.5	0.4	0.01	500	180	25
Impedance (Ω)	100	15	460	1000	10 000	0.05	1.3	25

2.
Current (A)	1.92	3.84	18.2	2.38	7.35	4.08	4.17	8.97
Volts V (a.c.)	4.7	7.5	225.7	230	107	228.5	400	235
Impedance (Ω)	2.45	1.95	12.4	96.3	14.56	56	96	26.2

3.
Impedance (Ω)	232	850	695.6	0.125	29.85	1050	129	4375
Volts V (a.c.)	176.3	230	400	26.5	0.194	457.8	238	245
Current (A)	0.76	0.27	0.575	212	0.0065	0.436	1.84	0.056

4. $101\,\Omega$ 5. $1.096\,A$ 6. (a) $2.18\,\Omega$, (b) $4.49\,\Omega$

7. (a) $0.472\,\Omega$, (b) $3.83\,\Omega$, (c) $0.321\,\Omega$, (d) $13\,\Omega$, (e) $0.413\,\Omega$ 8. $84.3\,\Omega$

9.
Volts V (a.c.)	61.1	105	153	193	230
Current (A)	2.3	4.2	6.12	7.35	9.2
Impedance (Ω)	26.56	25	25	26.26	25

10. (b)   11. (c)   12. (d)

Exercise 2

1. $4.71\,\Omega$ 2. $0.478\,H$

2.
Inductance (H)	0.04	0.159	0.12	0.008	0.152
Frequency (Hz)	50	50	48	90	60
Reactance (Ω)	12.57	50	36	4.5	57

4. (a) $40.8\,\Omega$, (b) $0.13\,H$ 5. (a) $15.97\,A$, (b) $13.07\,A$

6. (a) $3.77\,\Omega$, (b) $2.2\,\Omega$, (c) $0.141\,\Omega$, (d) $0.11\,\Omega$, (e) $14.1\,\Omega$

7. (a) $0.955\,H$, (b) $0.0796\,H$, (c) $0.0462\,H$, (d) $0.398\,H$, (e) $0.0159\,H$

10. $398\,V$ 11. (a)   12. (c)

Exercise 3

1. (a) $53\,\Omega$, (b) $127\,\Omega$, (c) $79.6\,\Omega$, (d) $21.2\,\Omega$, (e) $397\,\Omega$, (f) $265\,\Omega$, (g) $12.7\,\Omega$, (h) $33.5\,\Omega$, (i) $199\,\Omega$, (j) $42.4\,\Omega$

2. (a) $13.3\,\mu F$, (b) $42.4\,\mu F$, (c) $265\,\mu F$, (d) $703\,\mu F$, (e) $88.4\,\mu F$, (f) $199\,\mu F$, (g) $70.8\,\mu F$, (h) $7.96\,\mu F$, (i) $106\,\mu F$, (j) $44.2\,\mu F$

3. $346\,\mu F$ 4. $6.36\,\mu F$ 6. $159\,V$ 7. $207.6\,\mu F$ 8. $364\,V$

9. $10\,A$ 10. $15.2\,A$ 11. (d)   12. (a)

Exercise 4

1. R | 15 | 25 | 3.64 | 47.44 | 4.32 | 6.32 | 76.4 | 0.54
   R^2 | 225 | 625 | 13.25 | 2250 | 18.7 | 40 | 5837 | 0.735

2. X | 29.8 | 0.68 | 0.16 | 0.95 | 0.4 | 897 | 233.7 | 0.197
   X^2 | 888 | 0.46 | 0.026 | 0.9 | 0.16 | 804 609 | 54 616 | 0.039

3. 6.71 A   **4.** 8.69 A

5. $R(\Omega)$ | 14.5 | 140 | 9.63 | 3.5 | 57.6 | 94.8
   $X(\Omega)$ | 22.8 | 74.6 | 15.68 | 34.7 | 4050 | 49.6
   $Z(\Omega)$ | 27.02 | 159 | 18.4 | 34.87 | 4050 | 107

6. 232 Ω   **7.** 17.46 μF   **8.** (a) 16.9 Ω, (b) 73.3 Ω, (c) 71.3 Ω
9. 0.13 H, 115 V   **10.** (a) 28.75 Ω, (b) 0.122 H, (c) 47.9 Ω
11. 18.93 Ω, 15.04 Ω, 0.0479 H, 11.5 Ω   **13.** 69 μF
14. 0.318 H, 38.9 μF, 45.3 Hz   **15.** 14.57 A   **16.** (a) 7.47 A, (b) 127 μF
17. (c)   **18.** (c)

Exercise 5

1. 50 Ω   **2.** 40.1 Ω   **3.** 50 Ω   **4.** 198 Ω   **5.** 46.3 Ω
6. 231 Ω   **7.** 28 Ω   **8.** 1.09 Ω   **9.** 355 Ω   **10.** 751 Ω
11. 283 Ω   **12.** Approx. 500 Ω

13. Angle ϕ | 30° | 45° | 60° | 90° | 52° 24′ | 26° 42′ | 83° 12′ | 5° 36′
    sin ϕ | 0.5 | 0.7071 | 0.8660 | 1 | 0.7923 | 0.4493 | 0.9930 | 0.0976
    cos ϕ | 0.8616 | 0.7071 | 0.5 | 0.0 | 0.6101 | 0.8934 | 0.1184 | 0.9952
    tan ϕ | 0.5774 | 1 | 1.7321 | 0.0 | 1.2985 | 0.5029 | 8.3863 | 0.0981

14. Angle ϕ | 33° 3′ | 75° 21′ | 17° 15′ | 64° 29′ | 27° 56′ | 41° 53′
    sin ϕ | 0.5454 | 0.9675 | 0.2965 | 0.9025 | 0.4684 | 0.6676
    cos ϕ | 0.8382 | 0.2529 | 0.9550 | 0.4308 | 0.8835 | 0.7445
    tan ϕ | 0.6506 | 3.8254 | 0.3105 | 2.0949 | 0.5302 | 0.8967

15. Angle ϕ | 21° 48′ | 25° 48′ | 65° 30′ | 36° 52′ | 36° 52′ | 50° 24′ | 65° 20′ | 61° 36′
    sin ϕ | 0.3714 | 0.4352 | 0.91 | 0.6 | 0.6 | 0.7705 | 0.9088 | 0.8797
    cos ϕ | 0.9285 | 0.9003 | 0.4146 | 0.8 | 0.8 | 0.6374 | 0.4172 | 0.4754
    tan ϕ | 0.4000 | 0.4835 | 2.1948 | 0.75 | 0.75 | 1.2088 | 2.1778 | 1.8505

16. Angle ϕ | 75° 3′ | 64° 16′ | 5° 25′ | 38° 34′ | 29° 38′ | 72° 24′ | 72° 23′ | 71° 27′
    sin ϕ | 0.9661 | 0.9008 | 0.0946 | 0.6234 | 0.4945 | 0.9532 | 0.9531 | 0.9481
    cos ϕ | 0.2582 | 0.4341 | 0.9955 | 0.7819 | 0.8692 | 0.3020 | 0.3026 | 0.318
    tan ϕ | 3.7346 | 2.0752 | 0.0950 | 0.7973 | 0.5689 | 3.152 | 3.15 | 2.9814

17. 21.3 Ω, 20 Ω   **18.** 3.95 Ω, 6.13 Ω   **19.** 31° 47′   **20.** 90.9 Ω, 78 Ω
21. 191 W, 162 VAr   **22.** 32° 51′, 129 Ω   **23.** 28° 57′   **24.** 66.6 Ω
25. 37.6 Ω   **26.** 37.6

27. Phase angle ϕ | 75° 30′ | 72° 30′ | 65° 6′ | 60° | 56° 37′ | 53° 7′ | 45° 40′ | 34° 54′
    Power factor
    cos ϕ | 0.25 | 0.3 | 0.421 | 0.5 | 0.55 | 0.6 | 0.699 | 0.82

28. (a) 66.5 Ω, (b) 1.5 A, (c) 0.526, (d) 79.1 W
29. (a) 24.78 A, (b) 5700 W, (c) 0.86
30. (a) 17.8 Ω, (b) 0.03 H, (c) 12.92 A, (d) 0.844
31. 4 Ω, 6.928 Ω, 8 Ω

Exercise 6
1. 230 V   **2.** 31.1 A, 14.1 A   **4.** 151 V, 44° 30′   **5.** 3.63 A

Exercise 7
5. 248 V   **7.** 1029 Ω, 2754 Ω   **8.** 7 kW, 7.14 kVA   **9.** 2130 VA
10. 179 W   **11.** 5.1 A   **12.** 2.5 A   **13.** 197 V

Exercise 8
1. 0.47 A (lead)   **2.** 3.4 A, 27° 55′ (lag), 0.88 (lag)   **3.** 1.41 A (lag)
4. 2.78 A, 0.86 (lag)   **5.** 3.49 A, 0.92 (lag)   **6.** 10.6 μF   **7.** 1.71 A
8. (b)   **9.** (c)

Exercise 9
1. (a) 44.36 A, (b) 35.3 A, (c) 36.1 A, (d) 66 A, (e) 79.3 A, (f) 8.05 A, (g) 20.9 A,
(h) 59.1 A
2. 84%   **3.** 85.3%, 0.76   **4.** 76.9%, 0.754   **6.** 18.23 A   **7.** (d)   **8.** (a)

Exercise 10
1. 21.74 A; (a) 4 kW, (b) 3 kW   **2.** 131 kW, 141 kVA
3. 6.72 kW, 8.97 kVA, 356 μF   **4.** 11.5 kVA, 4.6 kW, 7.09 kVAr
5. 13.8 kVA, 6.9 kW, 8.6 kVAr, 37.39 A   **6.** 29.24 A
7. 124 μF, 5.35 A

8.

Power factor	0.7	0.75	0.8	0.85	0.9	0.95	1.0
Capacitance required μF	1137	1624	2109	2589	3092	3619	4825

9. 31.1 A; (a) 414 μF, (b) 239 μF   **10.** Approx. 15 μF   **11.** (b)   **12.** (a)
13. (b)   **14.** (c)

Exercise 11
1. 4.62 A, 2561 W   **2.** 1.63 A, 793 W
3. I_R = 9.2 A, I_Y = 17.73 A, I_B = 11.5 A; 5221 W   **4.** 9.58 A, 4962 W
5. (a) 23 A, (b) 41.6 A   **6.** I_R = 21 A, I_Y = 28.9 A, I_B = 24.9 A
7. (a) 7.7 A, 7.7 A, 5313 W, (b) 13.33 A, 23.1 A, 1600 W
8. 19.21 A, 7388 W   **9.** (a) 5.17 A, 2073 W, (b) 2.98 A, 1193 W
10. (a) 2.16 A, 0.469 (lag), 698 W, (b) 6.5 A, 0.469 (lag), 2190 W
11. (a) 7.17 Ω, (b) 21.51 Ω   **12.** (a) 884 μF, (b) 295 μF
13. I_{RY} = 6.66 A, I_{YB} = 8 A, I_{BR} = 12.6 A; 5422 W
14. (a) 6.09 kW, (b) 22.6 A   **15.** 7.14 kW, 14.72 A

16. (a) 17.2 A, (b) 29.8 A, (c) 20 900 W
17. (a) 11 kV, (b) primary $I_P = I_L = 400$ A, secondary $I_P = 0.09$ A, $I_L = 15.7$ A

Exercise 12
1. 40 mV   **2.** 45.2 Ω **3.** 99 975 Ω **4.** 1.5×10^{-3} Ω
5. 9960 Ω, 149 960 Ω, 249 990 Ω; 40×10^{-3} Ω, 4×10^{-3} Ω **6.** (c)
7. (d)   **8.** (a)   **9.** (b)

Exercise 13
1. (a) 15.6 lx, (b) 8 lx   **2.** 4.77 lx   **3.** 27 lx
4. 100 lx, 71.6 lx, 35.4 lx, 17.1 lx   **5.** 47.2 lx, 51.2 lx
6. (a) 56 lx, (b) new lamp of 385 cd or same lamp, new height 2.38 m
7. (c)   **8.** (a)

Exercise 14
1. 0.08 V   **2.** 2.56 m/s   **3.** 180 V   **4.** 102 V   **5.** 6.5 V   **6.** 0.15 H
7. 100 H   **8.** 1.75 H   **9.** 50 V   **10.** 66.7 A/S   **11.** 30 H   **12.** 80 V
13. (c)   **14.** (a)   **15.** (c)   **16.** (d)

Exercise 15
1. 183.72 A. Thermal storage is probably on its own installation, if the shower could be on its own control, then normal 100 A consumer unit can be used.
2. 113.26 A. Propose that the under sink heaters be on their own consumer unit.
3. 230.32 A. See paragraph 2 of *IEE On-site Guide*. This is a single-phase supply at present; consultation with the supplier would be essential. Perhaps a poly-phase supply would be available but could incur additional service cable costs.
4. 171 A (approx. 60 A per phase)

Exercise 16
1. 8.6 A   **2.** (a) 1.26 V, (b) 88.2 W   **3.** (a) 45 A, (b) 59.1 A, (c) 16 mm^2 (d) 4.9 V
4. (a) 13.04 A, (b) 7.2 V, (c) 5.86 V, (d) 224.14 V   **5.** (a) 75.76 A, (b) 1.96 V
6. (a) 2.22 mV, (b) 25 mm^2, (c) 4.05 V
7. (a) 52.2 A, (b) 60 A, (c) 100.44 A, (d) 1.69 mV/A/m, (e) 35 mm^2, (f) 4.4 V
8. (a) 43.5 A, (b) 50 A, (c) 64.73 A, (d) 0.96 mV/A/m, (e) 50 mm^2, (f) 4.85 V
9. (a) 28.58 A, (b) 32 A, (c) 34 A, (d) 5.95 mV/A/m, (e) 16 mm^2, (f) 4 V,
   (g) 32 mm conduit
10. (a) 39.13 A, (b) 4.47 mV/A/m, (c) 40 A, (d) 55.47 A, (e) 10 mm^2,
   (f) 229.14 V, 227.93 V, 226.56 V
11. (a) 21.74 A, (b) 25 A, (c) 25 A, (d) 19.2 mV/A/m, (e) 4.0 mm^2, (f) 4.18 V
12. (a) 27.95 A, (b) 9.2 V, (c) 30 A, (d) 30 A, (e) 8.22 mV/A/m, (f) 8.16 V
13. (a) 34.78 A, (b) 40 A, (c) 40 A, (d) 6.0 mm^2, (e) 4.44 V
14. (a) 42.45 A, (b) 63.7 A, (c) 96.8 A, (d) 10 mV/A/m, (e) 50 mm^2, (f) 0.34 V
15. (b)   **16.** (b)   **17.** (a)   **18.** (c)   **19.** (d)

Exercise 17

1. (a) 0.72 Ω, (b) 1.02 Ω   **2.** (a) 1.0 Ω, (b) 4.51 Ω

3. (a) 0.594 Ω, (b) 6.56 Ω   **4.** (a) 0.566 Ω, (b) 406 A, (c) 0.3 s

5. (c)   **6.** (c)   **7.** (b)

Exercise 18

1. 6 Nm

2.

Torque (Nm)	0	1.6	2.4	3.125	3.85	4.55
Power (W)	0	241.3	359	461	556	648

3. 795 N   **4.** 531 N, 177 N

5. (a) 112 Nm, (b) 13 Nm, (c) 229 Nm, (d) 1.61 Nm, (e) 31.8 Nm

6. 7.94 A

7.

Po (W)	0	33.2	66.0	98.4	130	161
n (%)	0	27.2	43.7	50.5	58.8	61.7

8. 1.4 A   **9.** 8.64 m/s   **10.** 9.05 m/s   **11.** 25.5 rev/s

12. (a) 229 rev/min, (b) 91.7 rev/min, (c) 76.4 rev/min, (d) 57.3 rev/min, (e) 153 rev/min

13. 78.8 mm   **14.** (a) 13.5 rev/s, (b) 750 rev/min

15. (a) 291 mm, (b) 181 mm   **16.** 3.61 A, 95 mm

Exercise 19

1. 430 V   **2.** (a) 90 V, (b) 126 V   **3.** (a) 442 V, (b) 84 A

4. (a) 100 A, (b) 257 V   **5.** 12.5 rev/s   **6.** (a) 32 Hz, (b) 24 Hz

7. 16.7 rev/s   **8.** 4.67%   **9.** 4%   **10.** 15.9 rev/s   **11.** 301.5 Nm

12. 25 000 MΩ

13. (a) 15 000 MΩ, (b) 7500 MΩ, (c) 3750 MΩ, (d) 2500 MΩ, (e) 1875 MΩ

14. 22.2 MΩ   **15.** 317 Ω   **16.** (b)   **17.** (b)   **18.** (b)